Design of Control Laws and State Observers for Fixed-Wing UAVs

Design of Control Laws and State Observers for Fixed-Wing UAVs

Simulation and Experimental Approaches

Arturo Tadeo Espinoza-Fraire
Faculty of Engineering, Science, and Architecture
University Juárez of the Durango State
Gómez Palacio, Durango, Mexico

Alejandro Enrique Dzul López
Electrical and Electronic Engineering Department
Technological Institute of La Laguna
Torreón, Coahuila, Mexico

Ricardo Pavel Parada Morado
Academy of Engineering in Manufacturing Technologies
Polytechnic University of Gómez Palacio
Gómez Palacio, Durango, Mexico

José Armando Sáenz Esqueda
Faculty of Engineering, Science, and Architecture
University Juárez of the Durango State
Gómez Palacio, Durango, Mexico

ELSEVIER

Elsevier
Radarweg 29, PO Box 211, 1000 AE Amsterdam, Netherlands
The Boulevard, Langford Lane, Kidlington, Oxford OX5 1GB, United Kingdom
50 Hampshire Street, 5th Floor, Cambridge, MA 02139, United States

ISBN: 978-0-323-95405-1

For information on all Elsevier publications
visit our website at https://www.elsevier.com/books-and-journals

Publisher: Matthew Deans
Acquisitions Editor: Giglio Chiara
Editorial Project Manager: Sara Greco
Production Project Manager: Prasanna Kalyanaraman
Cover Designer: Mark Rogers

Typeset by VTeX

Contents

List of figures

List of tables

Preface

This text is intended for use by undergraduate and postgraduate students and even researchers interested in UAVs. It can even be used in courses in aerodynamic engineering and automatic control theory.

We aim to achieve two objectives with this book. The first is the effectiveness to use a reduced uncoupled aerodynamic model for fixed-wing UAVs to design linear and nonlinear control laws and state observers. The second is to present several simulations (numerical computation) and experimental real flight test results with linear and nonlinear control laws and state observers designed from the reduced uncoupled aerodynamic model, which is helpful for any student or researcher due to the complexity of carrying out these tests with fixed-wing UAVs and even provide the information necessary to carry out real flight tests with such unmanned vehicles.

This book provides more than enough material for a one-semester course. The structure of this book is as follows:

- Chapter 1 presents a brief introduction to the classification of UAVs and nonmilitary applications of fixed-wing UAVs and a general description of control systems and state observers.
- Chapter 2 contains aerodynamic principles to understand the flight of fixed-wing UAVs.
- Chapter 3 is about obtaining the equations of motion of a fixed-wing UAV and contains the aerodynamic model used to design the nonlinear and linear control laws and state observers.
- Chapter 4 and Chapter 5 present the linear and nonlinear control laws, respectively. The results obtained with several computer simulations are presented.
- Chapter 6 is about the linear and nonlinear state observers and presents the results obtained with several computer simulations.
- Chapter 7 presents the real flight test results obtained with the linear and nonlinear control laws and with the state observers.
- The appendix presents essential material to understand the material presented in this book.

We recognize the support and help for this project from our families and Dennis McGonagle and Sara Greco from the Elsevier team.

A.T. Espinoza-Fraire

Acknowledgments

A.T. Espinoza-Fraire: To my wife Abril Huizar and my children Mariela and Mateo.

A. Dzul: To my wife Rosy and my children Jaime and Regina.

R. Pavel: To my wife Alma Lilia and my daughters Sophia and Maia.

J. Sáenz: To Dr. Campa, my father Dr. Agustín Sáaenz, and my mother María Concepción Esqueda.

Synopsis

Currently, the application of drones or UAVs presents a wide field of study for researchers to solve civil or industry problems. But to obtain solutions in these areas it is necessary to provide a basis for studying unmanned aerial systems.

This book provides the basis for studying fixed-wing UAVs, including a brief introduction into state-of-the-art fixed-wing UAVs. In addition, the development of mathematical models that describe these vehicles is included.

The controllers and state observers discussed in this book are linear and nonlinear. The results of several experiments (real flight tests) and simulations with the linear and nonlinear controllers and state observers are presented. The embedded system is given and the electronic components are described.

The appendix contains a mathematical review necessary to understand the linear and nonlinear theory used in the control laws and the state observers in this book. The appendix also contains the program codes of some linear and nonlinear controllers and observers used in this book, and it even includes the necessary program codes to read the sensors used in the embedded system. The MATLAB® code for the ground station to read the data from the fixed-wing unmanned aerial system during flight tests is added.

Introduction

1

This chapter presents a brief general overview of some classifications, applications, and control systems realized to conduct research and develop applications in the field of fixed-wing unmanned aerial vehicles (UAVs).

1.1 Classification of UAVs

UAVs can generally be classified into fixed-wing aircraft and multirotors. Both types of air vehicles possess unique characteristics that challenge their performance as autonomous systems. In this book, only UAVs of the fixed-wing class are considered.

The term "small UAV" refers to a class of fixed-wing UAVs with a wingspan between 5 and 10 feet. Small UAVs are usually fueled and require a runway for take-off and landing; in some cases, a catapult is used for take-off.

The term mini UAV (MAV) refers to a class of UAVs with a wingspan smaller than 5 feet. Small air vehicles are typically designed to fly from 10 to 12 hours and may load approximately 10 to 50 pounds [9]. Usually, these vehicles run on batteries, are launch-handed, and may not have landing gear, so they do not require a runway for take-off or landing.

MAVs are designed to fly from 20 minutes to a few hours; payloads range from ounces to several pounds. Consequently, the small payload severely restricts the load of sensors and microcontrollers that can be placed onto MAVs [9]. These restrictions pose exciting challenges for the design of autonomous systems.

1.2 Nonmilitary applications of fixed-wing UAVs

Nowadays, there exist many applications of UAVs. For example, applications of the fixed-wing UAV in agriculture include the detection of insects, weeds, diseases, crop characteristics, biomass, and background soil properties.

Marine applications include following species like whales and dolphins in order to analyze their movement and displacements across the ocean.

Other applications of UAVs include research of air quality. This is achieved by equipping the UAV with sensors to measure air pollution. Some sensors used in pollution analysis detect carbon dioxide, ozone, and black smoke particles.

Fixed-wing UAVs have already been applied to cartography, where it is necessary to interpret, analyze, and represent graphically all or part of a specific area, which is applied to mapping tasks.

Other interesting applications of this type of UAVs are in civil engineering and architecture, where they are used to obtain the topography of some terrain and con-

Design of Control Laws and State Observers for Fixed-Wing UAVs. https://doi.org/10.1016/B978-0-32-395405-1.00010-8

tour lines, for traffic analysis, for construction, in mining, and to analyze buildings' conditions.

There are many other applications of fixed-wing UAVs. In this section only a few nonmilitary applications are mentioned in order to have a general idea of this research field.

1.3 Control systems in fixed-wing UAVs

Besides having electronic and mechanical systems, UAVs need a mathematical algorithm processing the variables and constants necessary to achieve a specific task (dynamic modeling). This is where the technique or methodology of control enters. This section describes some control techniques applied to fixed-wing UAVs.

In [6] an algorithm to perform the tracking of a trajectory in the presence of unknown crossed wind is presented. This work is based on the adaptive backstepping technique. Some adaptation laws are proposed to estimate the unknown perturbation parameters used in the closed-loop system. In addition, a stability analysis is shown based on Lyapunov theory. An experimental platform is developed in [6] for the validation of the proposed controller.

The work [1] presents a controller by linearization to follow a path. This technique is applied to the dynamics of a fixed-wing UAV. In this work, a model of six degrees of freedom is used, where three degrees of freedom are for the Euler angles and the other three are for angular velocities. The feedback controller is combined with the open loop of the inverse dynamics of the plain, and then a closed loop is used for error regulation. The idea of applying open and closed loops is to consider perturbations and excitations of the internal dynamics. The feedback gains are obtained symbolically in the function of the nominal trajectory and thus avoid programming the gains.

In [26] an adaptive backstepping controller is used to control the longitudinal dynamics direction of a fixed-wing UAV. The designed nonlinear controller causes the system to follow the desired reference for the angle of incidence of the wind. In addition, a controller is designed for aerodynamic flight speed. These control objectives are achieved using the elevators and the front engine. In this work, a nonlinear model is used, and the proposed adaptation scheme allows them to design an adaptive controller with minimum knowledge of the aerodynamics of the planes.

Other techniques have been used for route or path tracking, such as the one employed in [51], which presents a method to follow a route exactly for fixed-wing MAVs. The procedure is based on the notion of vector fields, which are used to generate the desired inputs for the control laws designed for the position of the vehicle.

The vector field theory with the laws of control is developed for straight lines, circular arcs, and orbits paths. The Lyapunov approach is used to show asymptotic stability in path tracking. In addition, it is considered that the wind disturbances are constant.

In [2], an adaptive control law for an autopilot of a UAV is proposed. The adaptation algorithm offers the motion dynamics in a prescribed format for different flight conditions. On the basis of the simulation results, they verify the efficacy of the adaptive controller considering uncertainty in the parameters.

Another control technique used in fixed-wing UAVs is the linear quadratic regulator (LQR) control technique. In [71] the technique was applied to a UAV known as Flying Wing. This controller is used to stabilize the longitudinal movement of the airplane.

In [62], a procedure for designing an optimal LQR control is proposed. The LQR controller designed in [62] is applied to the longitudinal dynamics of a fixed-wing UAV. In [62] the open-loop dynamics and longitudinal flight modes are investigated. The design of the controller considers the open-loop time-domain response to suppress the damping system. In [71] and [62] the LQR controller is applied only to the longitudinal dynamics of a fixed-wing UAV.

To build an autopilot, the proportional-integral-derivative (PID) controller is used in [7]. The planning trajectory problems are another of the fields of interest in fixed-wing UAVs. In [43] optimal path planning algorithms are proposed in coordinated time for UAVs. For these trajectories, the fixed-wing UAVs must maintain a minimum speed at all times. Also, a simple computational algorithm for these vehicles that determines simultaneous trajectories from arbitrary starting points is presented.

In [28] two different control strategies are proposed, in which GPS data are utilized for the control of a flight. A comprehensive system was developed in that work, allowing the two control strategies to complement one another. The result is a more reliable flight control system to fly in areas with restricted GPS signals. Its performance has been analyzed through simulation and in-flight experiments.

In [15] a robust control is proposed for a fixed-wing UAV. The proposed controller is based on an adaptive control technique and high-order sliding mode controllers. The position and angular velocity of the vehicle are used for the design of the controller. Additionally, the moment of inertia is estimated. This control scheme increases robustness, and it is considered that it is not necessary to know the disturbance. The uncertainties of the model and external disturbances are taken into account.

Artificial vision systems are widely used in fixed-wing UAVs, as shown in [8], where a method to determine the GPS location of an object on the ground when the image has been taken from an MAV is presented. The main contribution of [8] is the presentation of four techniques to reduce the error of location. A filtering technique is discussed, and a path or flight path is chosen.

This book presents linear and nonlinear controllers applied to fixed-wing UAVs. The linear controllers are the proportional-derivative (PD), PID, LQR, and LQR with discrete-time Kalman filter.

The nonlinear controllers presented in this book are based on nested saturation, backstepping, and sliding mode control. In Fig. 1.1 a block diagram of the linear and nonlinear controllers which are used in this book is presented.

Finally, the MIT rule is proposed with the gradient method with sliding mode theory. Fig. 1.2 presents a block diagram of the adaptive control which is used in this book.

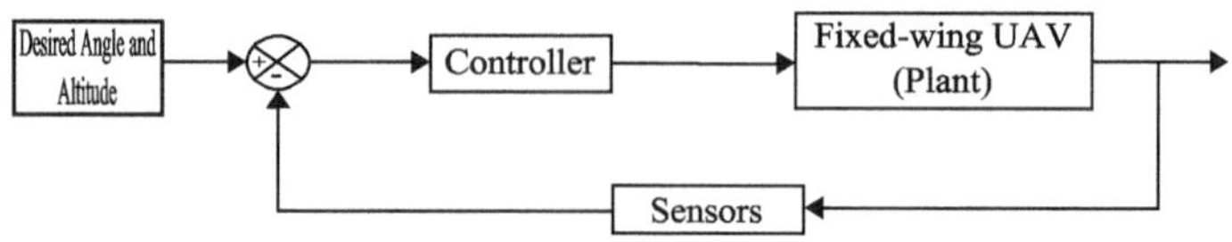

Figure 1.1 General control scheme.

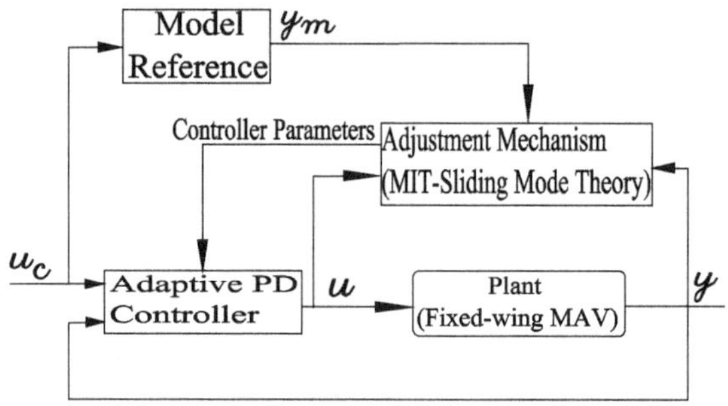

Figure 1.2 Block diagram of the MRAS applied to fixed-wing UAVs.

1.4 State observer systems in fixed-wing UAVs

Since the research of Luenberger published in 1966 [45], state observers have demonstrated good performance in automatic control for regulation tasks.

State observers have significant practical challenges because they are designed based on dynamic models; thus, the presence of disturbance, nonmodeling dynamics, and nonlinear characteristics reduces their performance. The design of robust state observers is pursued to improve performance, and some actual interesting progress has already been reported.

The sliding mode observer (SMO) was introduced by Slotine [59] and Utkin [64] in 1987. The robust state observer presented in [70] considers nonlinear bounded uncertainties, and the observer was designed based on the system variable structure method. The SMO presents a great improvement in performance in the presence of uncertainties and disturbances. In R. Sreedhar [53], the SMO (sliding mode observer) can be implemented a detect failure in nonlinear systems. In F. J. J. Hermans [30], a robust tracking control is reported, and the SMO is included in the control design. However, the asymptotic convergence between observer state and real state (i.e., observation error) is not guaranteed. Years later, an improvement ensured the exponential asymptotic convergence of the observation error [19].

J. Han proposes nonlinear extended state observer (NESO) in [31]. The NESO design is innovative because it does not require knowledge of the dynamical model. Thus, the designed observer is robust. The NESO was tested in the principal problems of industrial controls [32,24,69]. In [25], the authors use the algebraic solution of Riccati's equation to develop a new robust state observer for dynamic systems with perturbations. The observer guarantees the observation error approaches zero if it is possible to satisfy some conditions. In [5] this is demonstrated.

Time delays are a typical problem in automatic control. The time delays are consequences of variable sampling times, and they decrease the performance of the control law and observer. The time delays are presented in any implementation because the implementation is always with a digital computer. The computer operative system can in some cases be adjusted to reduce the time delays to low values, but in many other cases this is not possible, and we found time delays that affect the control performance. Hence, the presence of time delay is an important study topic. Some developments to address time delays are [20–22,57,58,11,66–68,12,54,23,17,4,61].

We will now discuss the research on state observers for linear systems. In [20], the authors present an observer for time delay and sufficient conditions obtained from linear matrix inequalities, and the reference [21] shows the same case for discrete time. A reduced-order observer for neutral linear systems is presented in [22] which proves stability despite time delay. In [57] a design for observers with unknown inputs is proposed.

The methodology presented in [58] solves the variable time delay case by SMO. However, they assume that the time delay and time delay limits are known. Based on such knowledge, the authors propose several kinds of state observers. Most of the time, delay reports require linear states or normal bounded linear states in the literature reviewed (see [67,68]).

This book presents linear and nonlinear observers applied to fixed-wing UAVs. The linear observer is the Luenberger observer, and the nonlinear observers are SMO, NESO, and the backstepping observer. In Fig. 1.3 a block diagram of the state observers which are used in this book is presented.

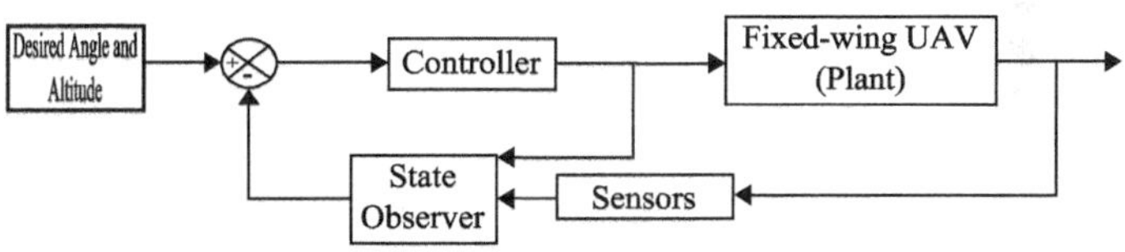

Figure 1.3 General state observers scheme.

Aerodynamic principles

2

The aerodynamic principles are essential for understanding the physics involved in fixed-wing UAVs. This chapter defines some basic definitions of aerodynamics. Section 2.1 presents the importance of knowledge of the aerodynamic principles, the definition of the atmosphere, atmospheric pressure, the standard atmosphere, air temperature, air density, airplane wings, the Bernoulli theorem, and the center of pressure. In Section 2.2 the forces acting in flight are shown, and definitions of flight opposition (resistance), thrust, and lift are provided. Section 2.3 presents the concepts of the axes of an airplane, aircraft control surfaces, and the structure of an aircraft.

2.1 The importance of aerodynamic principles

The dynamic model of an airplane usually has a complex mathematical definition, which involves existing aerodynamic aspects of the flight such as air humidity, airspeed, the pressure exerted on the aircraft, etc. These model parameters change as the aerial vehicle moves through the Earth's atmosphere.

Obtaining a dynamic model of a real aircraft is a fundamental process. An aircraft designer generates the required aircraft model from physical models and mathematical models that incorporate experimental data. The dynamic modeling process is often iterative. Thus, a mathematical model based on the laws of physics must be adjusted according to the data obtained in experimentation.

Due to the high cost of construction and flight tests, obtaining a suitable mathematical model of a real aircraft is of utmost importance since it is used to evaluate the aircraft's performance in combination with computer simulations. These models can also be used for pilot training in flight simulators, reconstructing flight conditions in the event of an accident, and studying the effects of modifications made to the design.

In addition, other mathematical models are used in all aspects of aircraft design (for example, structural models for the study of stress distribution and fatigue prediction). This book is a reference which can be used to design linear and nonlinear control laws and state observers. The definition and study of these controllers and observers will be presented in the chapters corresponding to these topics.

To obtain a suitable mathematical model, it is necessary to consider some basic concepts of aerodynamics, some of which will be presented below.

The basic concepts of aerodynamics are presented to explain, in a simple way and with basic terminology, why an airplane flies, that is, why an object heavier than air moves through this medium. The concepts of aerodynamic forces (lift, opposition, and thrust) given in this chapter are sufficient and elementary. We do not provide an exhaustive or detailed explanation of the complexities of aerodynamics.

Design of Control Laws and State Observers for Fixed-Wing UAVs. https://doi.org/10.1016/B978-0-32-395405-1.00011-X

2.1.1 The atmosphere

The atmosphere is the gaseous layer surrounding the earth, composed mainly of gases (78% nitrogen, 21% oxygen, and 1% other gases) that we call air. In the lower layers, the water vapor concentration must be added to these components, which depends on weather conditions and geographical location and can vary between 0% and 5%. As the water vapor concentration increases, the concentrations of the other gases decrease proportionally.

This gaseous element that we call air has many essential properties, but to explain why an airplane flies, the essential characteristics that define its behavior as a gas must be studied: pressure, temperature, and density. These three concepts are closely related and have a very significant effect on the flight behavior of an airplane [51].

2.1.2 Atmospheric pressure

Pressure is defined as the amount of force applied per unit area. According to this definition, atmospheric pressure is the force exerted by the atmosphere on a unit area, a force due to the weight of the air contained in an imaginary column.

Atmospheric pressure at a point represents the weight of a column of air of unit cross-sectional area extending from that point to the upper limit of the atmosphere. Since air density decreases with increasing height, this weight cannot be calculated unless we can express the variation in air density as a function of altitude or pressure. Therefore, it is not easy to calculate the atmospheric pressure on the Earth's surface.

This circumstance indicates a first air quality, which is a decrease in pressure at higher altitudes. The magnitude of this decrease is one millibar for every nine meters of height or approximately one inch for every thousand feet. Because of this property, aircraft flying above a certain altitude must be equipped with passenger cabin pressurization systems.

Among the instruments for measuring atmospheric pressure are the mercury barometer and the aneroid barometer. Some devices that make up the aircraft's instrumentation base their operation on this pressure reading. The units generally used in aviation are millibars or inches of mercury [51].

2.1.3 Standard atmosphere

The standard atmosphere, also known as type atmosphere or International Standard Atmosphere (ISA), is a hypothetical atmosphere based on average climatological measurements. Most important constants are based on the atmosphere at sea level.

Important constants:

- Temperature: 15°C (59°F)
- Pressure: 760 mm or 29.92 inches of mercury, equivalent to 1013.25 mb/cm^2
- Density: 1325 kg/m^3
- Acceleration due to gravity: 9.8 m/s^2

- Speed of sound: 340.29 m/s
- A thermal gradient of 1.98°C for each 1000 feet or 6.5°C for each 1000 meters
- A pressure drop of 1 inch for each 1000 feet, or 1 mb for each 9 meters, or 110 mb for each 1000 meters

This standard atmosphere, defined by the International Civil Aviation Organization (ICAO), is a reference pattern. However, a pilot will rarely have the occasion to fly in this standard atmosphere.

2.1.4 Air temperature

The Sun's heat passes through the atmosphere without significantly raising its temperature. Still, the earth absorbs this heat, raising its temperature, and gradually gives it to the layers of air in contact with it. The further away the air layers are from the earth, the less heat they receive from it. Because of this, the second quality of air is that the temperature decreases with height; that is, the higher the altitude, the lower the temperature.

The magnitude of this decrease is approximately 6.5°C per 1000 meters or 1.98°C per 1000 feet. These magnitudes are valid up to a height of 1100 meters or 3608.9 feet above sea level, after which the temperature is considered constant at −56.5°C. Although the magnitudes given are not met precisely, since air is not an ideal gas, these measured values are accepted as indicative of the behavior of air.

If a mass of gas contained in a container is heated, the pressure exerted by this mass on the container increases, showing a direct relationship between temperature and pressure. Thus, the pressure of warm air is greater than that of cold air [51].

2.1.5 Air density

The density of anything, solid, liquid, or gas, is expressed as its mass density per unit volume: $d = m/v$. The air density is low compared to water, but it is precisely this difference that makes flight possible.

If it is compressed, the same mass of gas will occupy less volume, or the same volume will hold more gas; that is, the density increases or decreases in direct relation to pressure. Since pressure decreases with height, this relationship also indicates that density will decrease; the higher the altitude, the lower the density. In addition, if heat is applied to a body, it expands and occupies more volume, so that in the same volume, there will be less mass or what is equivalent; its density will be lower. So as the temperature increases, the density of the air decreases.

A dilemma then arises because when increasing the height, on the one hand, the pressure decreases (density decreases), and on the other hand, the temperature decreases (density increases). Then, the pressure change has a more significant influence than the temperature change, resulting in a higher altitude with a lower density [51].

2.1.6 Airplane wing

The mathematical model used by control engineers usually contains data on the aircraft's aerodynamics as a whole. The study of the aerodynamic forces of the wing of an airplane is presented below. Fig. 2.1 shows the cross-section of a wing and defines its parts.

The flow field around the wing is represented by the cut lines shown in Fig. 2.2. The figure shows adjoining flow; the cut lines pass over the wing surface and do not change direction anywhere on the wing surface. Since the cross-section is uniform and the span length of this wing is assumed to be infinite, the flow field does not change in the direction perpendicular to the plane of the diagram. The free flow velocity vector defines the initial direction of the flow field. This is the speed measured ahead of the wing, far enough away that the flow is not affected by the presence of the airfoil.

The shape of the wing determines the aerodynamic properties and some essential geometric parameters. The line known as the chord length is a straight line drawn from the leading edge to the trailing edge and is the reference line used to describe the shape. A wing can be symmetric or, as usual, asymmetric concerning the chord.

The difference between the response line and the string line shows the amount of combat. The shape of the top and bottom surfaces, the amount of camber, the thickness, and the edge radius combine to determine the aerodynamic properties and useful speed range.

Two perpendicular components conventionally describe aerodynamic forces: the lift force and the drag force. The lift force Fl is defined perpendicular to the free flow of the velocity vector V, while the drag force is parallel to it. Lift and drag typically increase as α increases, where α is the angle of attack.

The expressions for the forces and moments generated by the airfoil depend on the lift coefficient, the drag coefficient, and the moment coefficients.

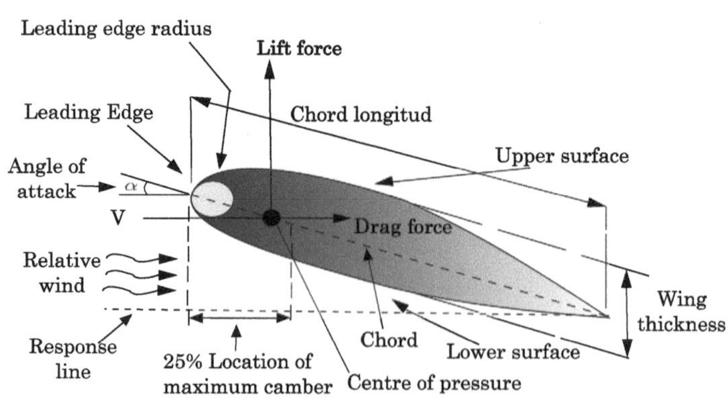

Figure 2.1 Schematic of an airfoil section.

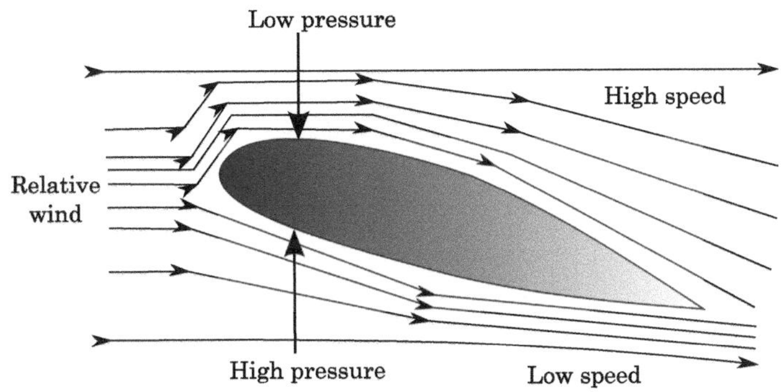

Figure 2.2 Airflow around the wing.

2.1.7 Bernoulli theorem

Daniel Bernoulli experimentally verified that the internal pressure of a fluid (liquid or gas) decreases as the velocity of the fluid increases; in other words, in a moving fluid, the sum of the pressure p and the velocity v in any point remains constant, that is, $p + v = k$. For this constant k to be maintained, if a particle increases its speed v, it will be at the cost of decreasing its pressure p.

Bernoulli's theorem is usually expressed in the form $p + \bar{q} = k$, where p denotes static pressure and $\bar{q} = \frac{1}{2}dv^2$ denotes the dynamic pressure, where d is the air flux density.

Air endowed with pressure p and density d flowing at speed v contains kinetic energy, just like any other moving object. According to the law of energy conservation, the sum of both is a constant. Given this equation, assuming that the air particles around the plane have the same density, then if the speed v increases, the pressure p will decrease, and vice versa [51].

2.1.8 The center of pressure

The center of pressure is the theoretical point of the wing where all the lift force is considered to be applied. Fig. 2.3 shows an example of pressure distribution on a profile moving in the air. For theoretical purposes, although the pressure acts on the entire airfoil, lift as the whole force is considered to be exerted on a point on the chord line.

The position of the center of pressure is usually expressed as a percentage of the chord of the wing from the leading edge. As the angle of attack increases or decreases, the pressure distribution around the profile changes, moving the center of pressure within certain limits, forwards or backward, respectively. The displacement margin is usually between 25% and 60% of the chord. Since it affects the stability of the aircraft, it is convenient that it be the smallest value possible; see Fig. 2.4.

The center of pressure tends to shift by moving the attack angle; when the attack angle is set to an angle where the pressure point comes out from between 25% and 60% of the chord, instead of ascending, the plane will tend to descend. This is when the wing is said to have stalled due to displacement of the center of pressure out of range before climbing the chord [14].

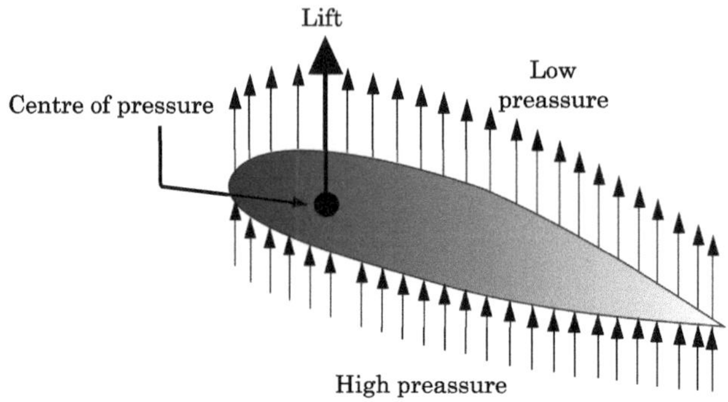

Figure 2.3 Center of pressure.

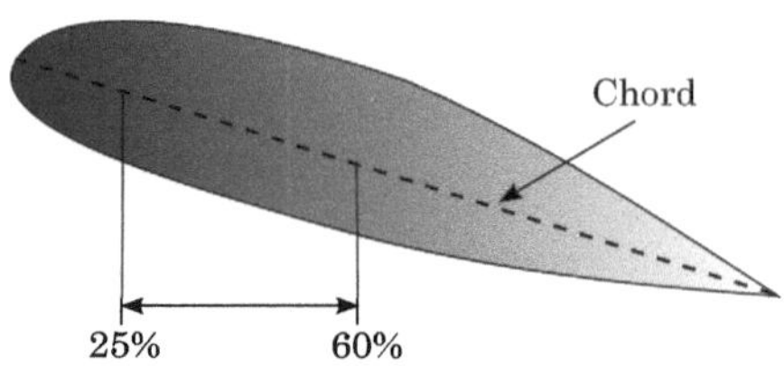

Figure 2.4 Pressure displacement limits.

2.2 Forces acting in flight

A series of forces act on an aircraft in flight that can be favorable and unfavorable for the aircraft's flight. Although experts continue to debate and investigate aerodynamics, for this book, we are only interested in some fundamental concepts such as those shown above to define the forces involved in a flight and their effects.

Of all the forces that act on an airplane in flight, the basic and main ones are lift, weight, thrust, and resistance; see Fig. 2.5. These forces are the main ones because all the maneuvers with an airplane are possible with these forces acting in flight. These four forces work in pairs: lift opposes weight and thrust opposes drag.

An airplane, like any object, remains stationary on the ground due to the action of two forces: its weight due to gravity that keeps it on the ground and the resistance to movement that supports its motionlessness. For this airplane to fly, it will be necessary to counteract the effects of these two forces, weight and opposition, using two other forces in the opposite direction, lift and thrust, respectively. Thus, the thrust has to overcome the resistance that opposes the plane's progress, and the lift must overcome the weight of the aircraft to keep it in the air [14].

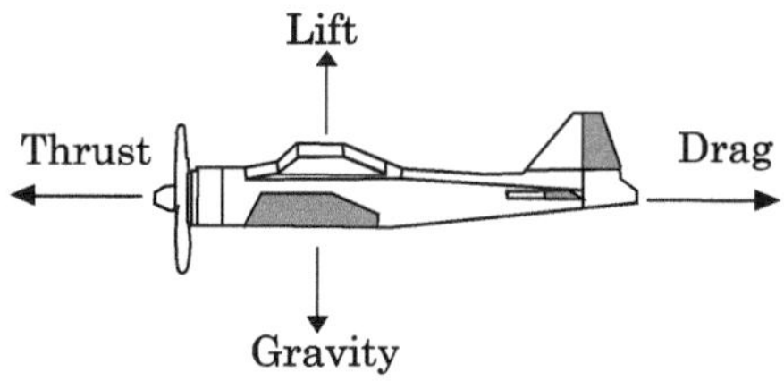

Figure 2.5 Forces acting in flight.

2.2.1 Flight opposition (resistance)

Drag is the force that prevents or opposes the motion of an aircraft. The resistance acts parallel and in the same direction as the relative wind. However, we could also say that the resistance is parallel and in the opposite direction to the advance towards the front of the plane. From an aerodynamic point of view, when a wing travels through the air, there are two forces of resistance: (a) the resistance due to the friction of the air on the wing and (b) the resistance due to the pressure of the air itself opposing it to the movement of an object.

The frictional resistance is proportional to the viscosity (opposition of fluid to tangential deformations), which in the air is very low. Frictional resistance is negligible, while pressure resistance depends on the mass density of air.

Both resistors create a force proportional to the area they act on and the speed square. A part of the pressure resistance produced by a wing depends on the amount of lift produced. This part is called induced resistance, and the sum of the rest of the resistances [14] is called parasitic resistance.

2.2.2 Thrust

To overcome the plane's position when it is not moving, it is necessary to accelerate it on the takeoff runway, maintain an adequate rate of climb, overcome resistance to advance, etc.; for this, a thrust or traction force is needed. This force is obtained by accelerating a mass of air at a higher speed than that of the airplane. The reaction, of equal intensity but in the opposite direction, moves the plane forward. In propeller aircraft, the propulsive force is generated by the propeller's rotation, driven by the conventional or turboprop engine. In the turbine, propulsion is achieved by the violent expulsion of the gases burned inside.

This force is exerted in the same direction as the axis of the propulsion system points, which is usually parallel to the aircraft's longitudinal axis. The main factor that influences this force is the engine's power, but other elements also influence it, such as the shape and size of the propeller, air density, etc.

Since power is equivalent to energy per unit of time, the higher the power, the greater the acceleration capacity. Aircraft power is the most critical factor in determining an aircraft's rate of climb. The maximum speed of climb of an airplane is not related to lifting but to the available power, [51].

2.2.3 Lift

Lift is the force generated on a body moving through a fluid in a direction perpendicular to the speed of the incident current. As with other aerodynamic forces, dimensionless coefficients represent the effectiveness of a body's shape in producing lift and are also used to facilitate calculations and designs.

It is the main force that allows a winged aircraft to stay in flight in aerodynamics. This, being more significant than the plane's total weight, will enable it to take off.

In addition, the lift, and consequently its coefficient, depends directly on the angle of attack, increasing as it increases until reaching a maximum point or a critical angle of attack, after which the airflow that passes over the upper surface (surface of the wing) does not manage to travel in its entirety and maintain adherence to the aerodynamic profile, giving rise to stall [14].

2.3 Axes of an airplane

The axes of an airplane are considered imaginary and ideal lines drawn on the aircraft, as shown in Fig. 2.6. Their denomination, as well as the movements that are made around them, are the following:

Longitudinal axis. This imaginary axis extends from the nose to the plane's tail. The movement around this axis is done by raising one wing and lowering the other; this movement is called roll.

Transverse or lateral axis. This imaginary axis is the one that extends from the end of one wing to the end of another wing. Movement is done by controlling the elevators to move the nose of the airplane and move it up or down. This movement is called pitching.

Vertical axis. This imaginary axis is the one that vertically crosses the center of the plane, and it is on which the nose (tip of the aircraft) turns to the left or the right. This movement is called yaw.

On a Cartesian coordinate axis, the longitudinal or roll axis would correspond to the x-axis; the transverse axis or pitch axis would be the y-axis; and the vertical axis or yaw axis would be the z-axis. The coordinate origin of this axis system is considered to be located at the plane's center of gravity [14].

2.3.1 Aircraft control surfaces

The plane's control is carried out using the aerodynamic surfaces that, activated by the pilot through the cockpit controls, modify the aircraft's aerodynamics, causing it to move on its axes. In this way, the trajectory tracks desired flight. Next, the primary control surfaces are mentioned, which are formed by ailerons, elevators, and stabilizers [14]. The flaps, slats, and spoilers are secondary control surfaces, as these are different surfaces some aircraft rely on to improve stability. The type of aircraft or fixed-wing UAV studied in this book does not have secondary control surfaces.

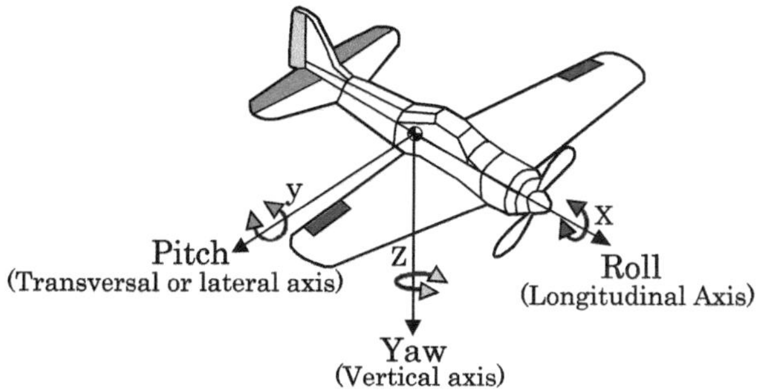

Figure 2.6 Axis of an airplane.

Ailerons. These are mobile surfaces located at the rear of the end of each wing, whose operation causes the aircraft to roll about its longitudinal axis. Its location is at the end of the wing because in this part the torque of the force exerted is greater. The pilot actuates the ailerons by turning the control wheel left or right.

The ailerons have an asymmetric movement. When turning the steering wheel to one side, the aileron of one wing goes up, and that of the opposite wing goes down, both at an angle of deflection proportional to the amount of turn given to the steering wheel. The aileron that rises on the wing, located towards where the steering wheel moves, implies less curvature in that part of the wing and therefore less lift, which causes that wing to go down, and the aileron below the opposite wing supposes greater curvature and lift, which makes that wing go up. This combination of opposing effects produces the rolling motion toward the descending wing.

Elevator. This is the mobile surface located at the rear of the horizontal empennage of the aircraft tail. Its activation causes the pitching movement of the aircraft on its transverse axis. The pitch movement of the plane causes the angle of attack to change, i.e., the elevator control knob controls the angle of attack. The elevator is actuated by the pilot pushing or pulling the steering wheel or control stick and typically has a maximum deflection of 40 degrees up and 20 degrees down.

Pulling the control drive raises this surface while pushing it lowers it. The rudder up produces less lift on the tail, which reduces the tail, and therefore, the nose goes up, thus resulting in a greater angle of attack. The rudder down increases the lift in the tail; it goes up, and consequently, the nose goes down, resulting in a lower angle of attack. In this way, the pitch movement of the aircraft is produced and, by extension, the angle of attack is modified.

Steering rudder. The movable surface is mounted on the back of the aircraft's vertical tail empennage. Its movement causes the yaw movement of the plane on its vertical axis. However, this does not make the aircraft turn; what makes the plane turn is the movement of the ailerons; the rudder is usually used to balance the forces in turns or to center the aircraft on the desired path. It typically has a maximum deflection of 30 degrees on each side. This surface is operated using pedals located on the floor of the cabin.

When pressing the right pedal, the rudder turns to the right, causing an aerodynamic reaction in the tail that makes it turn to the left; therefore, the nose of the plane yaws to the right. The opposite happens when you step on the left pedal, rudder to the left, tail to the right, and nose to the left.

Secondary control surfaces are mentioned below: flaps, slats, and spoilers. These always work in pairs and symmetrically; that is, the activation of the corresponding control causes the same movement down or up of the surfaces on both wings, except in the movements of the spoilers, which complement the ailerons affecting lift, airfoil shape, and wing surface. The fact that a surface works and not its symmetrical shape can be a severe drawback. Also, it has a speed limit; it can only work correctly at a range of speeds. Commercial aircraft accidents have occurred due to the negligent use of some of these surfaces in flight, which has led to improved designs, incorporating elements that prevent their activation at inappropriate speeds.

Flaps. The flaps are high-lift surfaces whose function is to increase the aircraft's lift when it flies at speeds lower than those for which the wing has been designed. Located at the rear inner part of the wings, they deflect downwards symmetrically, both at the same time at one or more angles, with which they change the curvature of the wing profile, which is more pronounced on the top surface and less pronounced on the upper surface. The wing's lower surface, the wing area, and the angle of incidence increase lift and drag.

The flaps are actuated from the cockpit, using a lever, an electrical system, or any other system with various degrees of trim (10 degrees, 15 degrees), corresponding to different lever positions or electrical switches, and they are not lowered or raised in all their draft at one stroke, but gradually. In general, larger deflections increase drag more than lift.

Slats. These are high-lift surfaces that act in a similar way to flaps. They are located at the bottom of the wing. When diverting, channeling a high-speed air current towards the upper surface increases lift, achieving greater angles of attack without stalling. It is generally used in large aircraft to increase lift in low-speed operations such as landings and take-offs, although there are light aircraft models that also have them.

In many cases, their use and retraction are done in an automated way; as long as the pressure exerted on them is sufficient, the slats will remain retracted, but when this sufficient pressure decreases to a certain level that is close to the stall speed, the slats deploy automatically. Due to the sudden increase or decrease in a lift at near-stall rates, more care should be taken when flying at low speeds in aircraft with these devices.

Spoilers. The purpose of these surfaces is to decrease the lift of the aircraft. It works mainly on jet planes that achieve high speeds and lose the aircraft's speed in flight and facilitate landing, helping to slow down on the ground. In some aircraft, it is used to supplement the ailerons for lateral control and in-flight turns.

In Fig. 2.7 the flaps, slats, and spoilers of the wing of an airplane are shown.

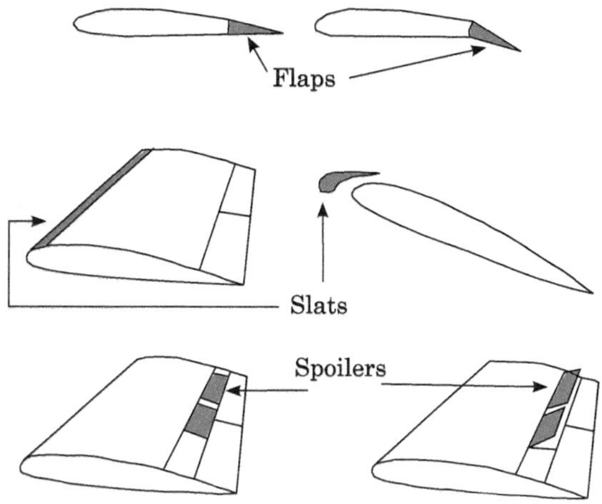

Figure 2.7 Location of flaps, slats, and spoilers on a wing.

2.3.2 The structure of an airplane

The aircraft parts mentioned below are necessary to carry out the control of the aircraft; see Fig. 2.8.

Fuselage. The fuselage is the name given to the main body of the aircraft structure, whose primary function is to house the crew, passengers, and cargo and serve as the primary support for the rest of the components. The airframe design must also provide acceptable performance for the aircraft's intended purpose. The fuselages that offer less aerodynamic resistance are those with a circular, elliptical, or oval section, also characterized by their elongated and conical shape [50].

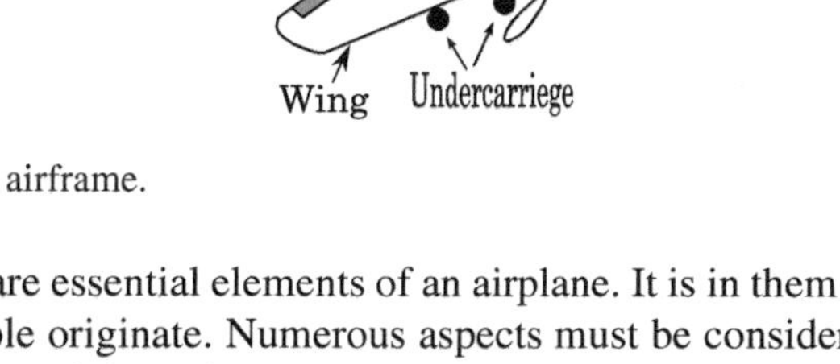

Figure 2.8 Generic airframe.

Wings. These are essential elements of an airplane. It is in them that the forces that make flight possible originate. Numerous aspects must be considered in their design, such as maximum weight to be supported, resistance generated, behavior when the wing stalls due to a change in pressure point, etc. Through these aspects, the aircraft aims to provide the optimum performance to combine the best speed with the greatest range and the lowest possible fuel consumption.

Stabilizer system. This is generally composed of a vertical stabilizer and a horizontal one (the elevator). As their names indicate, their mission is to contribute to the stability of the plane on its vertical and horizontal axes.

Landing gear. Its mission is to cushion the plane's impact on landing and allow the aircraft to roll and move on land. It can be fixed or retractable, and tricycle designs exist, that is, two main wheels and one wheel placed in the lower front part of the plane. Other types, such as the so-called tail skid, have two main wheels and a skid or tail wheel. There are snow-adapted trains with skids and trains adapted with floats to land on water.

Engine-propulsion group. This is responsible for providing the necessary power to counteract the device's resistance, both on the ground and in flight, to drive the wings, produce lift, and give the required acceleration. This group can be made up of one or more engines that can be a piston, reaction, turboprop, etc. This group can also include propellers, which can have different sizes, shapes, and numbers of blades.

Auxiliary systems. The rest of the systems serve as support for the operation of the previous elements or to provide more comfort or better control of the aircraft. We can mention, for example, the hydraulic system, the electrical system, the one in charge of pressurization, fuel supply, etc.

2.4 Concluding remarks

This chapter has presented fundamental concepts about aerodynamics that are necessary to understand the theory presented in this book. With these concepts, the reader can understand the flight of fixed-wing UAVs. It should be mentioned that these concepts are also applied to airplanes with pilots on board, not only to UAVs.

Equations of motion of a fixed-wing UAV

3

The aerodynamic model that defines a fixed-wing airplane can describe the dynamics of a fixed-wing UAV; this is because the fixed-wing UAV is flying in the same environment as a commercial or military aircraft. The UAV is perturbed by the same weather conditions in a real flight (wind gusts, temperature variations, etc.). Thus, this chapter presents a reduced aerodynamic model to obtain a mathematical model that does not need technology to calculate the aerodynamic parameters besides having a dynamic model that can be used to design various control laws.

This chapter introduces the aerodynamic model that defines a fixed-wing UAV. Section 3.1 presents the principal control surfaces of a fixed-wing UAV. Section 3.2 shows frame coordinates in the fixed-wing UAV. Next, Section 3.3 deals with the variables of a fixed-wing UAV, and in Section 3.4 the motion of a rigid body is shown.

Section 3.5 presents the kinematic model of the fixed-wing UAV. Section 3.6 deals with the uncoupled model of the fixed-wing UAV, presenting three reduced dynamics: longitudinal dynamics (Section 3.6.1), directional dynamics (yaw angle) (Section 3.6.2), and lateral dynamics (roll angle) (Section 3.6.2). Finally, Section 3.6.3 presents the change of variables of the reduced aerodynamic model.

3.1 Control surfaces of a fixed-wing MAV

The control variables of a fixed-wing MAV are the energy of the frontal motor and the usual or primary control surfaces denoted by the variable δ. The control surfaces perform the function of diverting the airflow and therefore produce aerodynamic forces and moments [50].

A fixed-wing UAV with a traditional configuration like in this book is composed of a propulsion motor. The elevator moves the pitch angle; this surface is defined as δ_e. The ailerons control the roll angle; this surface is defined as δ_a. The rudder governs the yaw angle; this surface is defined as δ_r. Fig. 3.1 shows the control surfaces.

3.2 Frame coordinates in fixed-wing UAVs

The coordinate frames often used to describe the movement of a UAV are the Earth-centered inertial reference frame E and the body-fixed frame B. In the Earth-centered inertial reference frame, the axis directions are north–east–down; see Fig. 3.2.

This frame is helpful to describe the position and orientation of the UAV. In the UAV's body-fixed frame, the origin is positioned in the center of gravity and has a

Design of Control Laws and State Observers for Fixed-Wing UAVs. https://doi.org/10.1016/B978-0-32-395405-1.00012-1

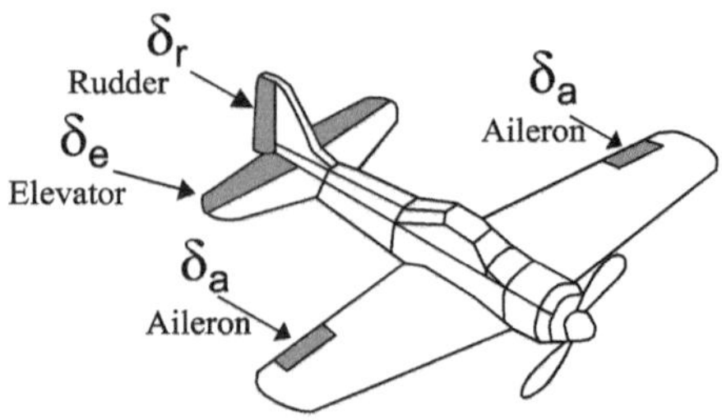

Figure 3.1 Control surfaces.

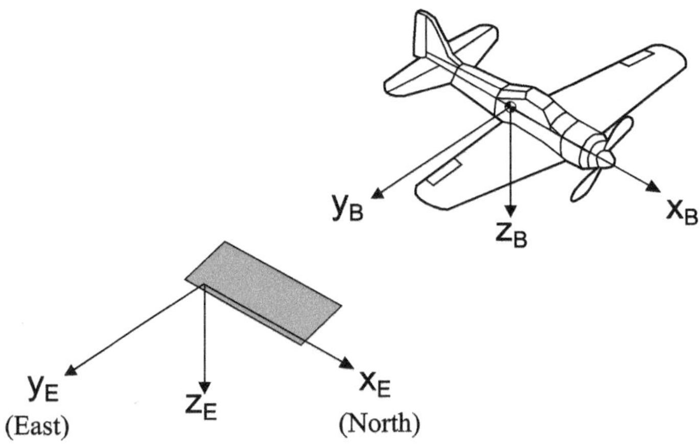

Figure 3.2 Fixed coordinate frame E and frame attached to the airplane B.

vector pointing toward the front of the airplane, another vector toward the right wing, and the last vector pointing toward down [50]. In this frame, the matrix of inertia moments of the MAV is constant; therefore, it is adequate to describe the rotation angles.

3.3 Governing physics of a fixed-wing UAV

Consider the airplane as a rigid body. The position can describe its motion, orientation, linear velocity, and angular velocity; see Fig. 3.3. For the airplane dynamics, it is considered a rigid body, and any flexible structure is omitted. The Earth's curvature is insignificant for distances that the UAVs fly. As an inertial system, therefore, it is possible to apply Newton's second law.

- $\mathbf{p} = (p_N \;\; p_E \;\; h)^T$. The vector $\mathbf{p}$ is the position of the airplane implicit in the Earth-centered inertial reference frame, where p_N is the coordinate in the north direction, p_E is the coordinate in the east direction, and h is the airplane's altitude.

- $\mathbf{v} = (u \quad v \quad w)^T$. The vector $\mathbf{v}$ is the linear velocity in the coordinate axes of the airplane, where the components u, v, and w represent the axial velocity, the lateral velocity, and the normal velocity in the body-fixed frame, respectively.
- $\mathbf{\Phi} = (\phi \ \theta \ \psi)^T$. The symbol $\mathbf{\Phi}$ describes the Euler angles, that is, the orientation of the airplane relative to the Earth-centered inertial reference frame, where ϕ is the roll angle, θ represents the pitch angle, and ψ is the yaw angle. These angles are related to the body-fixed frame with the Earth-centered inertial reference frame.
- $\omega = (p \quad q \quad r)^T$. The vector ω describes the angular velocities in the body-fixed frame, where p, q, and r are the angular velocity components in the directions of the x-, y-, and z-axes in the body-fixed frame, respectively.

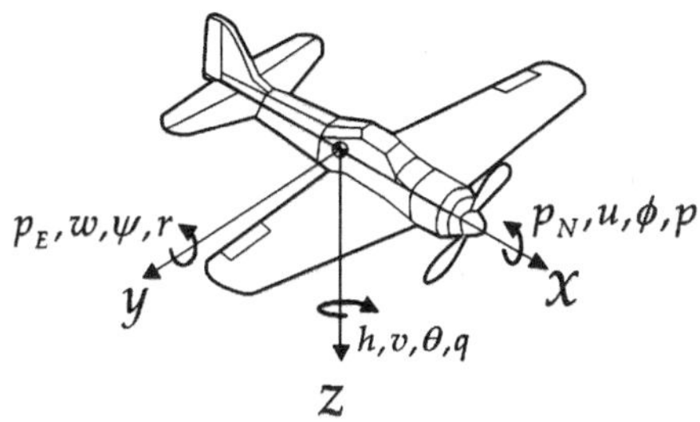

Figure 3.3 Movement variables of a fixed-wing UAV.

3.4 Motion of a rigid body

As mentioned before, the UAV is considered as a rigid body to analyze the force effects and the external moments; thereby, Newton's laws of motion can be applied [50]. Taking the body-fixed frame B, with Newton's second law, we obtain the following equations for a rigid body in terms of the linear velocity $\mathbf{v}$ and the angular velocity ω:

$$\mathbf{F} = m(\dot{\mathbf{v}} + \omega \times \mathbf{v}), \tag{3.1}$$

$$\mathbf{T} = \dot{\mathbf{H}} + \omega \times \mathbf{H}, \tag{3.2}$$

where $\mathbf{F}$ and $\mathbf{T}$ are respectively the total external force and the total external moment applied to the airplane, m is the mass, and $\mathbf{H}$ is the angular moment of the aircraft. Then, the angular moment is defined as

$$\mathbf{H} = I\omega, \tag{3.3}$$

$$I = \begin{pmatrix} I_x & 0 & -I_{xz} \\ 0 & I_y & 0 \\ -I_{xz} & 0 & I_z \end{pmatrix},$$

where I_{ii} denotes an inertia moment and I_{ij} is a product inertia, $j \neq i$; the zeros in the matrix I are because in the body-fixed frame the xz-plane is a plane of symmetry for most aircraft. The forces and moments are due to gravity, thrust, and the aerodynamic effects [50]. These are introduced in the total force equation **F** and in the entire moment equation **T**. From this fixed frame about the airplane the moments of inertia (matrix I) are constant.

The engine produces the thrust force, while the aerodynamic forces and moments are due to the interaction between the aircraft body and the surrounding airflow. By development of the force equation (3.1), we obtain

$$\begin{aligned}\mathbf{F} &= m(\dot{\mathbf{v}} + \omega \times \mathbf{v}) \\ &= m \begin{pmatrix} \dot{u} \\ \dot{v} \\ \dot{w} \end{pmatrix} + \begin{pmatrix} 0 & -r & q \\ r & 0 & -p \\ -q & p & 0 \end{pmatrix} \begin{pmatrix} u \\ v \\ w \end{pmatrix} \\ &= m \begin{pmatrix} \dot{u} \\ \dot{v} \\ \dot{w} \end{pmatrix} + \begin{pmatrix} -rv + qw \\ ru - pw \\ -qu + pv \end{pmatrix}. \end{aligned} \tag{3.4}$$

On the other hand, the force **F** can be written as

$$\begin{aligned}\mathbf{F} &= \begin{pmatrix} F_x \\ F_y \\ F_z \end{pmatrix} + \begin{pmatrix} F_{g_x} \\ F_{g_y} \\ F_{g_z} \end{pmatrix} \\ &= \begin{pmatrix} F_x \\ F_y \\ F_z \end{pmatrix} + mg \begin{pmatrix} -\sin\theta \\ \cos\theta \sin\phi \\ \cos\theta \cos\phi \end{pmatrix}. \end{aligned} \tag{3.5}$$

Using Eqs. (3.4) and (3.5), the resulting equations are

$$\begin{aligned} F_x - mg\sin\theta &= m(\dot{u} - rv + qw), \\ F_y + mg\cos\theta\sin\phi &= m(\dot{v} + ru - pw), \\ F_z + mg\cos\theta\cos\phi &= m(\dot{w} - qu + pv). \end{aligned} \tag{3.6}$$

Finally, by isolating the components of the angular accelerations the equations defined in [56] are obtained:

$$\begin{aligned} \dot{u} &= rv - qw - g\sin\theta + \frac{F_x}{m}, \\ \dot{v} &= -ru + pw + g\sin\phi\cos\theta + \frac{F_y}{m}, \\ \dot{w} &= qu - pv + g\cos\phi\cos\theta + \frac{F_z}{m}. \end{aligned} \tag{3.7}$$

The moment equation (3.2) is

$$\mathbf{T} = I\dot{\omega} + w \times I\omega, \tag{3.8}$$

where

$$I\dot{\omega} = \begin{pmatrix} I_x & 0 & -I_{xz} \\ 0 & I_y & 0 \\ -I_{xz} & 0 & I_z \end{pmatrix} \cdot \begin{pmatrix} \dot{p} \\ \dot{q} \\ \dot{r} \end{pmatrix} = \begin{pmatrix} \dot{p}I_x - \dot{r}I_{xz} \\ \dot{q}I_y \\ -\dot{p}I_{xz} + \dot{r}I_z \end{pmatrix} \tag{3.9}$$

$$\omega \times I\omega = \begin{pmatrix} 0 & -r & q \\ r & 0 & -p \\ -q & p & 0 \end{pmatrix} \cdot \begin{pmatrix} pI_x - rI_{xz} \\ qI_y \\ -pI_{xz} + rI_z \end{pmatrix} = \begin{pmatrix} -rqI_y - q(pI_{xz} + rI_z) \\ r(pI_x - rI_{xz}) + p(pI_{xz} + rI_z) \\ -q(pI_x - rI_{xz}) + pqI_y \end{pmatrix} \tag{3.10}$$

with **T** as the vector of external moments, whose components are

$$\mathbf{T} = \begin{pmatrix} \bar{L} \\ M \\ N \end{pmatrix}, \tag{3.11}$$

where $\bar{L}$, M, and N represent the roll moment, the pitch moment, and the yaw moment, respectively. Then, to obtain the applied moments in roll, pitch, and yaw, we have to add up (3.9) and (3.10), that is,

$$\begin{aligned} \bar{L} &= I_x\dot{p} - I_{xz}\dot{r} + qr(I_z - I_y) - I_{xz}pq, \\ M &= I_y\dot{q} + pr(I_x - I_z) - I_{xz}(p^2 - r^2), \\ N &= I_z\dot{r} - I_{xz}\dot{p} + pq(I_y - I_x) + I_{xz}qr. \end{aligned} \tag{3.12}$$

Additional aerodynamic forces are involved in the airplane's flight like the lift force L, the force of opposition D to the longitudinal movement in the x-axis of the airplane, and the lateral force Y. Generally, these aerodynamic forces are represented in the aircraft's airframe; it is worth mentioning that these aerodynamic forces are not considered in this book to design the control laws and state observers. Then, the mathematical expressions for the aerodynamic forces are

$$D = \bar{q}SC_D, \tag{3.13}$$

$$\bar{Y} = \bar{q}SC_Y, \tag{3.14}$$

$$L = \bar{q}SC_L, \tag{3.15}$$

where $\bar{q}$ is the aerodynamic pressure:

$$\bar{q} = \frac{1}{2}\rho V_T^2, \tag{3.16}$$

where ρ is the air density, V_T^2 denotes the wind velocity, and S represents the wing area of the airplane. The aerodynamic coefficients are C_D, C_Y, and C_L. The aerodynamic coefficients significantly affect the airplane or UAV dynamics; it is noteworthy that these coefficients are difficult to model numerically, but they can be calculated by a wind tunnel or computed using software simulations. Commonly, every coefficient is defined as a sum of various components.

A common approach consists of storing them in tables to later perform interpolations to determine intermediate values. Another approximation consists in trying to adjust the data to some parametric function. In Fig. 3.4 the moments and the aerodynamic forces in the airplane are presented.

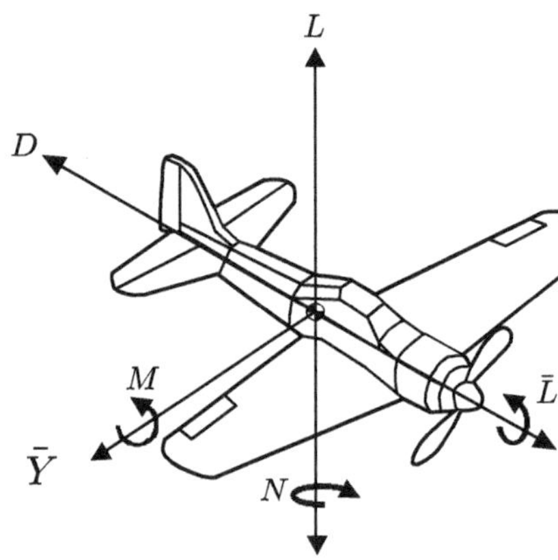

Figure 3.4 Moments and aerodynamic forces in the airplane.

3.5 Kinematic model

The orientation of the rigid body can be defined considering a coordinated system fixed in the rigid body. It is therefore necessary to specify the relationship between an inertial axis system and one attached to the body. It is possible to build the complete transformation composed of three rotations and three translations. A lot of conventions of Euler angles exist, given in the order of the rotations.

This book uses the convention whose sequence is the following: the inertial coordinate frame is rotated about the z-axis, corresponding to the angle ψ, and then this new axis system is rotated around the y-axis, and with this an angle θ is obtained to produce another intermediate set of axes. Finally, it is rotated at an angle ϕ about the x-axis.

All rotations are in a sense counterclockwise [56]. Each rotation takes place about an axis whose location depends on the previous rotation. As the three rotations occur about the z-, y-, and x-axes, this representation is known as Euler angles z-y-x (see Fig. 3.5) [13]. Suppose that it is necessary to transform a given vector in inertial coor-

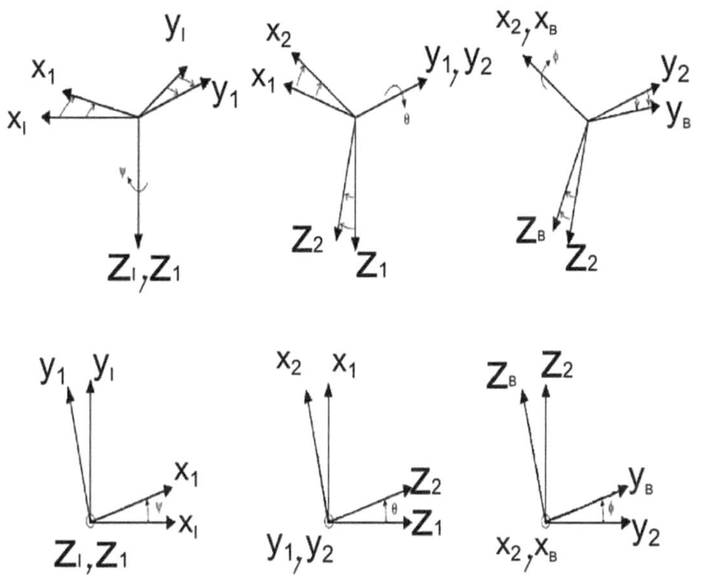

Figure 3.5 Representation z-y-x of Euler angles.

dinates (x_I, y_I, z_I) and a given vector in coordinates of the mobile frame (x_B, y_B, z_B). It can be performed using the previous rotation in Euler angles; the vector given in inertial coordinates is expressed in an intermediate coordinate frame (x_1, y_1, z_1). This intermediate frame is obtained by the first rotation of the inertial frame (x_I, y_I, z_I) and represents a rotation about the z_I-axis by angle ψ. The corresponding rotation matrix is

$$\mathbf{R}_{1I} = \begin{pmatrix} \cos\psi & \sin\psi & 0 \\ -\sin\psi & \cos\psi & 0 \\ 0 & 0 & 1 \end{pmatrix}. \tag{3.17}$$

The following rotation matrix indicates a rotation about the y_I-axis at angle θ:

$$\mathbf{R}_{21} = \begin{pmatrix} \cos\theta & 0 & -\sin\theta \\ 0 & 1 & 0 \\ \sin\theta & 0 & \cos\theta \end{pmatrix}. \tag{3.18}$$

Finally, the rotation matrix corresponding to a rotation about x_I at an angle ϕ is given as

$$\mathbf{R}_{B2} = \begin{pmatrix} 1 & 0 & 0 \\ 0 & \cos\phi & \sin\phi \\ 0 & -\sin\phi & \cos\phi \end{pmatrix}. \tag{3.19}$$

The complete transformation, from the inertial frame to the airplane frame, is given by the composition of the three transformations:

$$\mathbf{R}_{BI} = \mathbf{R}_{B2}\mathbf{R}_{21}\mathbf{R}_{1I} \tag{3.20}$$

$$= \begin{pmatrix} \cos\theta\cos\psi & \cos\theta\sin\psi & -\sin\theta \\ \cos\psi\sin\theta\sin\phi - \cos\phi\sin\psi & \cos\phi\cos\psi + \sin\theta\sin\phi\sin\psi & \cos\theta\sin\phi \\ \cos\psi\sin\theta\cos\phi + \sin\phi\sin\psi & -\sin\phi\cos\psi + \sin\theta\cos\phi\sin\psi & \cos\theta\cos\phi \end{pmatrix}. \tag{3.21}$$

For the inverse transformation the following equation is used:

$$\mathbf{R}_{BI} = \mathbf{R}_{IB}^{-1} = \mathbf{R}_{IB}^{T}. \tag{3.22}$$

To calculate the angular velocity from Euler angle derivatives it is necessary to apply the following procedure: it is assumed that the rotation occurs independently of each axis and the contributions of the rotations are added. The component of the angular velocity due to rotation at angle $\dot{\psi}$ about z is

$$\mathbf{R}_{BI}(\dot{\psi}z_I) = \mathbf{R}_{BI}\begin{pmatrix} 0 \\ 0 \\ \dot{\psi} \end{pmatrix}_I = \dot{\psi}\begin{pmatrix} -\sin\theta \\ \cos\theta\sin\phi \\ \cos\theta\cos\phi \end{pmatrix}_B. \tag{3.23}$$

The component due to rotation about the y_1-axis is

$$\mathbf{R}_{BI}(\dot{\theta}y_1) = (\mathbf{R}_\phi\mathbf{R}_\theta)(\dot{\theta}y_1) = \mathbf{R}_\phi(\dot{\theta}y_2)$$

$$= \begin{pmatrix} \cos\theta & 0 & -\sin\phi \\ \sin\phi\sin\theta & \cos\phi & \sin\phi\cos\theta \\ \cos\phi\sin\theta & -\sin\phi & \cos\phi\cos\theta \end{pmatrix}\begin{pmatrix} 0 \\ \dot{\theta} \\ 0 \end{pmatrix}_2 = \dot{\theta}\begin{pmatrix} 0 \\ \cos\phi \\ -\sin\phi \end{pmatrix}_B. \tag{3.24}$$

Finally, the component of the angular velocity due to rotation about the x_2-axis is given by

$$\mathbf{R}_\phi(\dot{\phi}x_2) = (\dot{\phi}x_2)\begin{pmatrix} 1 \\ 0 \\ 0 \end{pmatrix}_B. \tag{3.25}$$

Then, the angular velocity of the airplane defined in the airplane frame is given as

$$\omega = \begin{pmatrix} p \\ q \\ r \end{pmatrix}_B = \dot{\phi}\begin{pmatrix} 1 \\ 0 \\ 0 \end{pmatrix}_B + \dot{\theta}\begin{pmatrix} 0 \\ \cos\phi \\ \sin\phi \end{pmatrix}_B + \dot{\psi}\begin{pmatrix} -\sin\theta \\ \cos\theta\sin\phi \\ \cos\theta\cos\phi \end{pmatrix}_B$$

$$= \begin{pmatrix} 1 & 0 & -\sin\theta \\ 0 & \cos\phi & \cos\theta\sin\phi \\ 0 & -\sin\phi & \cos\theta\cos\phi \end{pmatrix}\begin{pmatrix} \dot{\phi} \\ \dot{\theta} \\ \dot{\psi} \end{pmatrix}. \tag{3.26}$$

The matrix which maps the derivatives from the Euler angles to the angular velocity of the airplane is known as $\mathbf{L}_{BI}$, that is,

$$\mathbf{L}_{BI} = \begin{pmatrix} 1 & 0 & -\sin\theta \\ 0 & \cos\phi & \cos\theta\sin\phi \\ 0 & -\sin\phi & \cos\theta\cos\phi \end{pmatrix}. \tag{3.27}$$

The determinant of $\mathbf{L}_{BI}$ is

$$\det(\mathbf{L}_{BI}) = \cos\theta. \tag{3.28}$$

The relationship between the angular velocities of the aircraft and the angular velocities of the Euler angles be inverted if $\theta \neq \pm\frac{\pi}{2}$. Assuming this is the case, it is defined that

$$\begin{pmatrix} \dot{\phi} \\ \dot{\theta} \\ \dot{\psi} \end{pmatrix} = \mathbf{L}_{BI}\boldsymbol{w}, \tag{3.29}$$

where

$$\mathbf{L}_{BI} = \mathbf{L}_{BI}^{-1} = \begin{pmatrix} 1 & \tan\theta\sin\phi & \tan\theta\cos\phi \\ 0 & \cos\phi & -\sin\phi \\ 0 & \frac{\sin\phi}{\cos\theta} & \frac{\cos\phi}{\cos\theta} \end{pmatrix}. \tag{3.30}$$

Then the operation (3.29) is as follows:

$$\begin{pmatrix} \dot{\phi} \\ \dot{\theta} \\ \dot{\psi} \end{pmatrix} = \begin{pmatrix} 1 & \tan\theta\sin\phi & \tan\theta\cos\phi \\ 0 & \cos\phi & -\sin\phi \\ 0 & \frac{\sin\phi}{\cos\theta} & \frac{\cos\phi}{\cos\theta} \end{pmatrix} \begin{pmatrix} p \\ q \\ r \end{pmatrix}. \tag{3.31}$$

Finally, performing operations in (3.31), the kinematic equations are

$$\dot{\phi} = p + \tan\theta(q\sin\phi + r\cos\phi), \tag{3.32}$$

$$\dot{\theta} = q\cos\phi - r\sin\phi, \tag{3.33}$$

$$\dot{\psi} = \frac{q\sin\phi + r\cos\phi}{\cos\theta}. \tag{3.34}$$

In summary, we can represent the equations defining the aerodynamics of an airplane or fixed-wing UAV developed above as follows.

Force equations:

$$\dot{u} = rv - qw - g\sin\theta + \frac{F_x}{m} \tag{3.35}$$

$$\dot{v} = -ru + pw + g\sin\phi\cos\theta + \frac{F_y}{m} \tag{3.36}$$

$$\dot{w} = qu - pv + g\cos\phi\cos\theta + \frac{F_z}{m} \tag{3.37}$$

Kinematic equations:

$$\dot{\phi} = p + \tan\theta(q\sin\phi + r\cos\phi) \tag{3.38}$$

$$\dot{\theta} = q\cos\phi - r\sin\phi \tag{3.39}$$

$$\dot{\psi} = \frac{q\sin\phi + r\cos\phi}{\cos\theta} \tag{3.40}$$

Moment equations:

$$\bar{L} = I_x\dot{p} - I_{xz}\dot{r} + qr(I_z - I_y) - I_{xz}pq \tag{3.41}$$

$$M = I_y\dot{q} + pr(I_x - I_z) - I_{xz}(p^2 - r^2) \tag{3.42}$$

$$N = I_z\dot{r} - I_{xz}\dot{p} + pq(I_y - I_x) + I_{xz}qr \tag{3.43}$$

Navigation equations:

$$\begin{aligned}\dot{p}_N = \dot{x} =& u\cos\theta\cos\psi + v(-\cos\phi\sin\psi + \sin\phi\sin\theta\cos\psi)\\ &+ w(\sin\phi\sin\psi + \cos\phi\sin\theta\cos\psi)\end{aligned} \tag{3.44}$$

$$\begin{aligned}\dot{p}_E = \dot{y} =& u\cos\theta\sin\psi + v(\cos\phi\cos\psi + \sin\phi\sin\theta\sin\psi)\\ & w(-\sin\phi\cos\phi + \cos\phi\sin\theta\sin\psi)\end{aligned} \tag{3.45}$$

$$\dot{h} = \dot{z} = u\sin\theta - v\sin\phi\cos\theta - w\cos\phi\cos\theta \tag{3.46}$$

3.6 Uncoupled model of the fixed-wing UAV

The obtention of the equations that describe an airplane's aerodynamics has been presented in the previous section. In this book, for the design of the control laws for a fixed-wing UAV, the aerodynamic model has been considered as uncoupled dynamics, considering the concepts defined in [50] and [56].

In this part, we obtain three models, describing the longitudinal dynamics, the dynamics of the yaw angle, and the dynamics of the roll angle of the fixed-wing UAV.

3.6.1 *Longitudinal dynamics*

In the longitudinal dynamics, a pilot can control the passage of the airplane and the reaction velocity by using the control surface called the elevator and the speed at which the aircraft moves on this axis. In this action, the elevator can change the pitch angle and the longitudinal velocity of the aircraft.

Some relationships can be analyzed to predict aircraft handling and are determined by the elevator control surface, the space where the airplane is flying (flight area), the angle of attack, the altitude, and the velocity. The techniques used by aircraft designers to establish the response of the aircraft can also be used by the designer of a controller or to develop a flight simulator to validate the dynamics of the simulated aircraft.

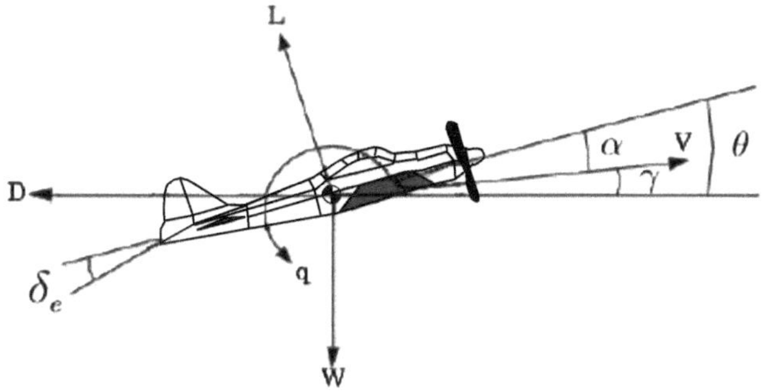

Figure 3.6 Longitudinal movement.

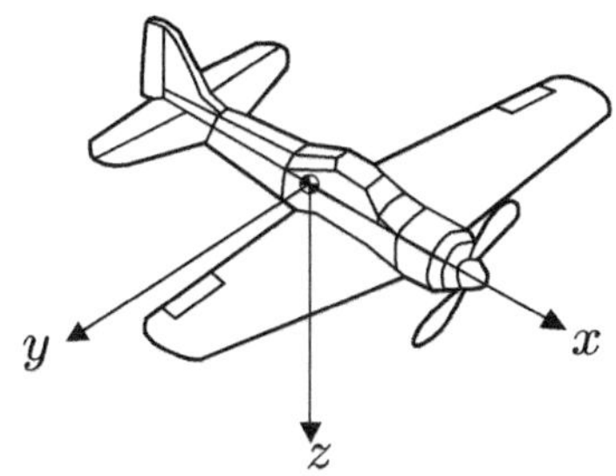

Figure 3.7 The x-, y-, and z-axes in the body of the airplane.

An interesting point in fixed-wing air vehicles is that, in general, the longitudinal dynamic can be separated from the lateral dynamic (roll) and the directional dynamic (yaw) to analyze separately the performance of the dynamic system [14].

This simplification is valid for an aircraft in which the geometry is symmetrical concerning the xz-plane of body axes (see Fig. 3.7) and we can assume that in the perpendicular thrust of the motor, the contributions of the angles of yaw and roll are insignificant so we can ignore these angular contributions. In this case, roll angle ϕ and yaw ψ are zero, and the lateral forces are equal to zero.

Although this simplification helps analyze the dynamics of fixed-wing UAVs, for the case of a rotary-wing aircraft, the effects of the rotor on the aircraft are somewhat difficult to analyze with a dynamic in this form.

The parameters involved in the dynamic model describing the longitudinal movement of a plane [56] are shown in Fig. 3.6. This model allows us to analyze the forward movement and the altitude control. The movements of rotation on the x-axis and the z-axis are called lateral and directional movements and for this case they are not considered. Fig. 3.9 and Fig. 3.8 present the lateral and directional movements of an airplane on the x-axis and the z-axis, respectively.

The airplane's altitude is related to its angle of inclination, which indicates whether the airplane's nose is pointing up or down. The pitch angle θ is the angle between the aircraft's horizon and the x-axis. Usually, a plane ascends gently so that the aerodynamics of its wings produce enough lift to overcome gravity. Therefore, the model for designing controllers for the altitude of the airplane is given by the following first-order differential equations:

$$\dot{\theta} = q, \tag{3.47}$$
$$\dot{q} = M_q q + M_{\delta_e} \delta_e, \tag{3.48}$$
$$\dot{h} = V \sin(\theta), \tag{3.49}$$

where:

θ is the pitch angle,
q is the component of the angular velocity about the y-axis on the body of the airplane,
h is the airplane altitude,
δ_e is the movement of the elevator (horizontal stabilizer) control surface on the airplane,
M_q is the derivative of stability,
M_{δ_e} is the derivative of control, and
V is the magnitude of the airplane speed.

The aerodynamic stability derivatives are defined by

$$M_q = \frac{\rho S V \bar{c}^2}{4 I_y} C_{m_q}, \tag{3.50}$$
$$M_{\delta_e} = \frac{\rho V^2 S \bar{c}}{2 I_y} C_{m_{\delta_e}}, \tag{3.51}$$

where:

ρ is the air density (1.05 kg/m^3),
S is the wing area (0.09 m^2),
$\bar{c}$ is the mean aerodynamic chord (0.14 m),
b is the wingspan (0.914 m),
I_{yy} is the y-component of the diagonal of the inertia matrix (0.17 kg·m^2),
C_{m_q} is the dimensionless coefficient for the stability of the longitudinal movement (-50), and
$C_{m_{\delta_e}}$ is the dimensionless coefficient for the stability of the elevator control (-0.25).

3.6.2 Directional and lateral dynamics

The directional and lateral dynamics of the airplane are assumed to be independent of the airplane's longitudinal dynamics. This simplification is helpful for the design of the controllers and state observers for the yaw and roll angles.

The lateral movement of the aircraft for turning maneuvers causes a change in velocity and altitude. However, the lateral dynamics can enter a new problem related to the roll angle and may induce a yaw rotation and vice versa; therefore, there is a coupling between the rotations on roll and yaw axes. It is assumed to analyze the equations of motion that the difference in these variables will be minimal and, therefore, may be omitted. We do not consider the effects of engine thrust on the design of the control laws and state observers in this book.

The objective of simplifying further analysis and facilitating the design of controllers is to consider a decoupling of the directional and lateral dynamics of the airplane. Thus, in an experimental test to follow a desired path with the yaw angle, it can be considered as a perturbation of the movement of the roll angle, proving in this way the robustness of the control laws and the state observers in a real flight test [50].

Directional dynamics (yaw angle)

The yaw angle is controlled through the control surface called a rudder (vertical stabilizer), which is a surface handled by the pilot. The following equations represent the yaw angle [56], [50]:

$$\dot{\psi} = r, \tag{3.52}$$

$$\dot{r} = N_r r + N_{\delta_r} \delta_r, \tag{3.53}$$

where:

ψ is the yaw angle,
r is the yaw angular velocity in reference to the center of gravity,
δ_r is the rudder deviation,
N_r is the derivative of stability, and
N_{δ_r} is the derivative of control.

The following formulas describe the derivatives of aerodynamic stability for yaw angle [50]:

$$N_r = \frac{\rho V S b^2}{4 I_z} C_{n_r}, \tag{3.54}$$

$$N_{\delta_r} = \frac{\rho V^2 S b}{2 I_z} C_{n_{\delta_r}}, \tag{3.55}$$

where:

I_z represents inertia on the z-axis (0.02 kg·m^2),
C_{n_r} is the dimensionless coefficient of damping in yaw angle (-0.01), and
$C_{n_{\delta r}}$ is the dimensionless coefficient of affectivity in the rudder (0.0005).

Fig. 3.8 presents the variables of the dynamic of the yaw angle.

Lateral dynamics (roll angle)

The ailerons generally control the roll angle, but these angles usually are too nonlinear and are often accompanied by a yaw angle [50]. Considering the natural relationship between the roll and yaw angles, the yaw angle in experimental tests for bank angle is considered a perturbation, thereby testing the robustness of the roll angle controller

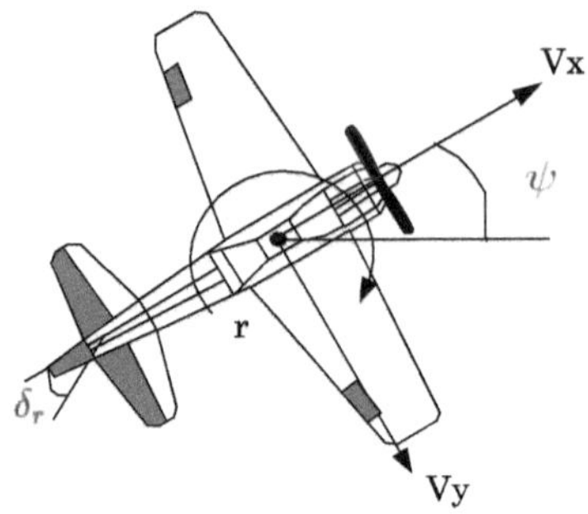

Figure 3.8 Yaw angle.

and the state observer. The equations describing the dynamics of roll angle are uncoupled from equations of yaw and of the longitudinal dynamics. Thus, the dynamics of the roll angle are

$$\dot{\phi} = p, \tag{3.56}$$

$$\dot{p} = L_p p + L_{\delta_a}\delta_a, \tag{3.57}$$

where:

ϕ is the roll angle,
p is the roll angular velocity in reference to the center of gravity,
δ_a is the deviation of the control surface known in aviation as ailerons,
L_p is the derivative of stability, and
L_{δ_a} is the derivative of control.

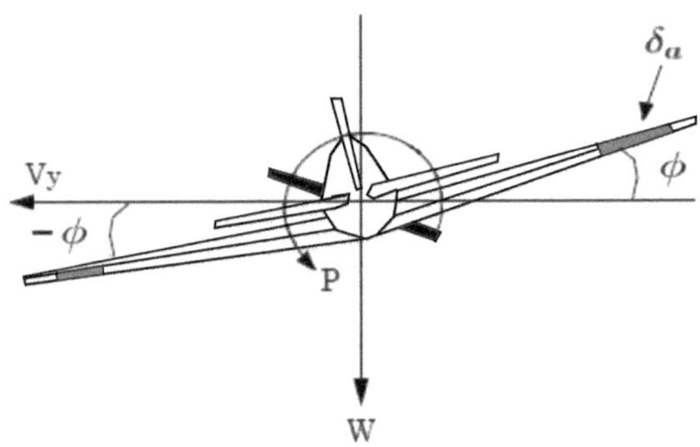

Figure 3.9 Roll angle.

In Fig. 3.9 the variables involved in the roll angle are shown. The derivatives of stability for the roll angle are defined by the following formulas [50]:

$$L_p = \frac{\rho V S b^2}{4 I_x} C_{l_p}, \tag{3.58}$$

$$L_{\delta_a} = \frac{\rho V^2 Sb}{2I_x} C_{l_{\delta_a}}, \tag{3.59}$$

where:

I_x is the inertia on the x-axis (0.16 kg·m^2),
C_{lp} is the dimensionless coefficient of damping in the roll angle (−0.15), and
$C_{l_{\delta_a}}$ is the dimensionless coefficient of affectivity in ailerons (0.005).

3.6.3 Change of variables

In the design of linear and nonlinear controllers and the state observers in the subsequent chapters, we have defined a change in the variables notation for the dynamics that describe the movement uncoupled in altitude, yaw, and roll angles. Because the dynamics are similar, with this change of variables it is possible to avoid repeating each control law and state observer. Then, the change of variables in the longitudinal dynamics (3.47)–(3.49) is given by

$$\dot{x}_{1\theta} = x_{2\theta}, \tag{3.60}$$
$$\dot{x}_{2\theta} = C_{1\theta} x_{2\theta} + C_{2\theta} u_\theta, \tag{3.61}$$
$$\dot{h} = V \sin(x_{1\theta}). \tag{3.62}$$

The change of variables for the dynamics (3.52)–(3.53) of the yaw angle is defined by

$$\dot{x}_{1\psi} = x_{2\psi}, \tag{3.63}$$
$$\dot{x}_{2\psi} = C_{1\psi} x_{2\psi} + C_{2\psi} u_\psi. \tag{3.64}$$

Finally, the change of variables corresponding to the dynamics of the roll angle (3.56)–(3.57) is

$$\dot{x}_{1\phi} = x_{2\phi}, \tag{3.65}$$
$$\dot{x}_{2\phi} = C_{1\phi} x_{2\phi} + C_{2\phi} u_\phi. \tag{3.66}$$

3.7 Concluding remarks

In this chapter we applied a Newton–Euler approach to obtain the aerodynamic model that defines a fixed-wing UAV. The complete mathematical model is separated to describe the pure pitching motion, the pure yaw and roll motions to design different control techniques, state observers for the altitude movement, and the yaw and roll angles. But the model conserves the most critical aerodynamic derivatives defined in the aerodynamic theory, which are the stability and control derivatives.

The mathematical model obtained in this chapter will be used in later chapters to design control laws and state observers and to develop the game theory applied to fixed-wing UAVs.

Linear controllers

4

In the last years, the study and analysis of control laws to stabilize UAVs have been the branches to explore in automatic control theory.

This chapter proposes linear control laws to achieve a stable flight with a fixed-wing UAV; this means maintaining a fixed altitude, achieving a roll angle of zero degrees, and moving the yaw angle into angular values that do not produce singularities.

This chapter is organized as follows. In Section 4.1 the PD and PID controllers are presented, which are classic control laws. In Section 4.2 the LQR is developed and Section 4.3 is devoted to an LQR with discrete-time Kalman filter.

The controllers are designed with the decoupled models of the longitudinal dynamics, yaw, and roll, respectively. These dynamic models were presented in Section 3.6 of Chapter 3.

This chapter presents the simulation results obtained with the linear controllers mentioned above.

4.1 PD and PID controllers

For the design of PD and PID controllers in altitude, the error $\tilde{e}_h = h_d - h$ is defined, where h_d is the desired altitude and h is the current altitude of the fixed-wing UAV which is obtained with the integral of Eq. (3.62). Altitude control is achieved by varying the pitch angle θ; then, for varying θ an error of the form $\tilde{e}_\theta = \theta_d - \theta$ is defined, where $\theta_d = \arctan(\tilde{e}_h/\varsigma)$ is the desired pitch angle, where ς is the longitude from the center of mass to the nose of the fixed-wing UAV.

For the control of the yaw angle, an error $\tilde{e}_\psi = \psi_d - \psi$ is defined, where ψ_d is the desired yaw angle and ψ is obtained from the integral of Eq. (3.63). We have to define an error to achieve the desired roll angle, which is defined as $\tilde{e}_\phi = \phi_d - \phi$, where ϕ_d is the desired roll angle and ϕ is the obtained roll angle by integrating Eq. (3.65). Thus, the PD controller obtained is given by [52]

$$u_l = k_{pl}\tilde{e}_l + k_{vl}\dot{\tilde{e}}_l, \tag{4.1}$$

where k_{pl} and k_{vl} are positive gains. To know which movement is being applied in the control law, a subscript l is used, that is, $l = \theta, \psi, \phi$; this subscript will be used for the linear and nonlinear controllers. The PID controller [52] is defined by

$$u_l = k_{pl}\tilde{e}_l + k_{il}\int_0^t \tilde{e}_l(t)dt + k_{vl}\dot{\tilde{e}}_l, \tag{4.2}$$

where k_{pl}, k_{il}, and k_{vl} are the gains called proportional, integral, and derivative respectively; these gains are positive definite [42]. Table 4.1 presents the gain values for the PD controller used in the simulation.

Design of Control Laws and State Observers for Fixed-Wing UAVs. https://doi.org/10.1016/B978-0-32-395405-1.00013-3

Table 4.1 PD controller gains.

Movement	k_{pl}	k_{vl} [s]
Altitude	10	8
Yaw	4	3.5
Roll	2	2.5

Table 4.2 presents the values of the PD controller gains. The controller's PD and PID parameters are adjusted to obtain an overdamped response, and further, the current signal reaches the reference signal in less than 10 seconds. Also, the control objectives mentioned above were achieved with a control action that was not very big. The control action determines the deflection of the control surfaces of the fixed-wing UAV. In the case of the altitude control, the elevator deflection δ_e is used. For the yaw and roll angles, the deflection is the rudder (δ_r) and ailerons (δ_a), respectively. For the change of variables described in Section 3.6.3, $\delta_e = u_\theta$, $\delta_r = u_\psi$, and $\delta_a = u_\phi$.

The values used in the simulation gains for the linear and nonlinear controllers and observers are obtained in a heuristic form. For the experiments, we started the real flight test with the gains used in the simulations, and after the first test, it is necessary to retune the gains to obtain the desired flight.

Table 4.2 PID controller gains.

Movement	k_{pl}	k_{il} [1/s]	k_{vl} [s]
Altitude	10	5	0.025
Yaw	4	3.5	0.0001
Roll	2	2.5	0.0001

In Fig. 4.1 the altitude simulation results obtained with the PD controller are presented. It is observed that the fixed-wing UAV achieved the desired altitude within 8 seconds (see the upper graph of Fig. 4.1). The control action applied by the PD controller on altitude is shown in the lower graph of Fig. 4.1.

Fig. 4.2 shows the altitude simulation results for the PID controller; one can appreciate that this controller achieved the desired value within 8 seconds in the same way as the PD controller for this movement (see the upper graph of Fig. 4.1). The lower graph of Fig. 4.2 shows the control action of the PID on altitude.

For the yaw angle simulation results of the linear and nonlinear controllers, a non-constant signal is proposed, which changes its value at a specific time.

For the particular case of the simulations presented, the reference angle of the yaw starts with 1.9 degrees, then at 30 seconds it changes to 2 degrees, then at 60 seconds it changes to 5 degrees, and at 90 seconds it changes to 2 degrees. This reference is proposed because with a fixed altitude and with a roll angle of 0 degrees, the yaw angle can take different reference values without the fixed-wing UAV becoming unstable.

The results obtained with the PD controller for the yaw angle are shown in Fig. 4.3, where a small impulse in the yaw angle of the fixed-wing UAV is observed (see the upper graph of Fig. 4.3). In the lower graph of Fig. 4.3, the control action applied by the rudder using the PD control law can be observed.

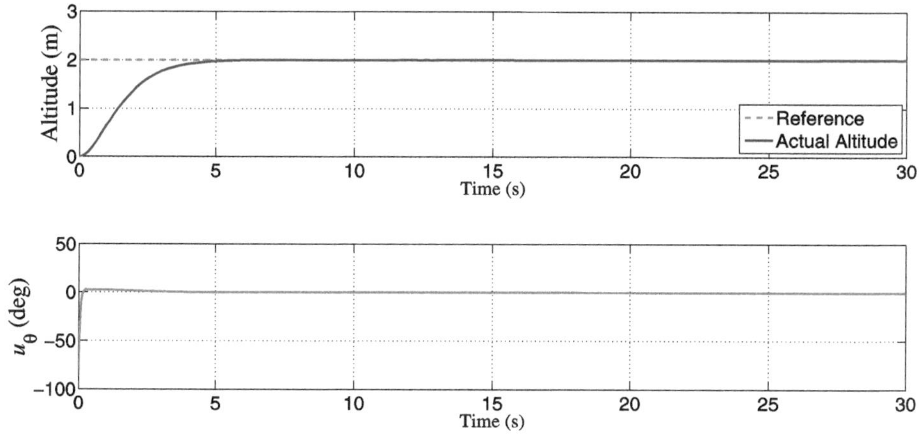

Figure 4.1 Action of the PD controller on altitude.

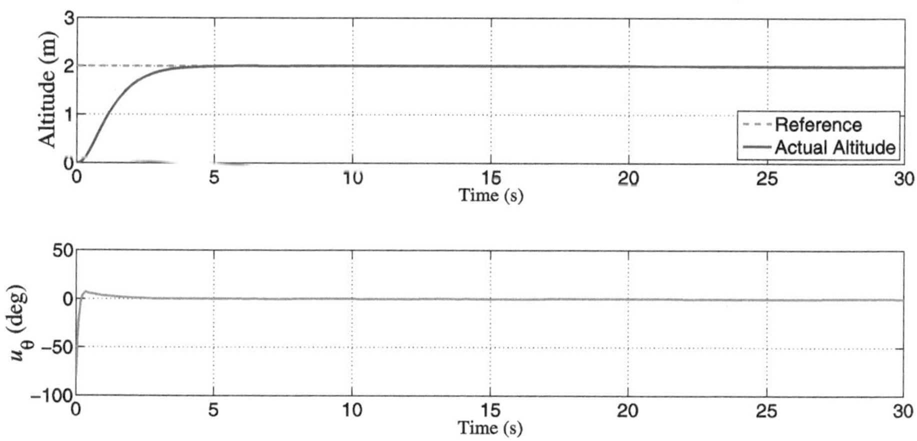

Figure 4.2 Action of the PID controller on altitude.

In Fig. 4.4 the yaw results with the PID controller are shown; the same reference signal as for the PD is considered (see the upper graph of Fig. 4.4).

In the lower graph of Fig. 4.4, the control action generated by the PID controller on yaw is shown. Fig. 4.5 shows the action of the PD controller on the roll. For the roll control, it is considered in the simulation that the fixed-wing UAV has an initial roll angle of 5 degrees, and the controller has to lead the fixed-wing UAV to a final roll angle of 0 degrees and keep it there.

In the upper graph of Fig. 4.5, it is observed that the fixed-wing UAV achieved a roll angle of 0 degrees in less than 10 seconds. The lower graph of Fig. 4.5 shows the action of the PD controller on the roll.

The simulation results with the PID controller are shown in Fig. 4.6. This simulation considered the same reference signal and the same initial condition as for the PD

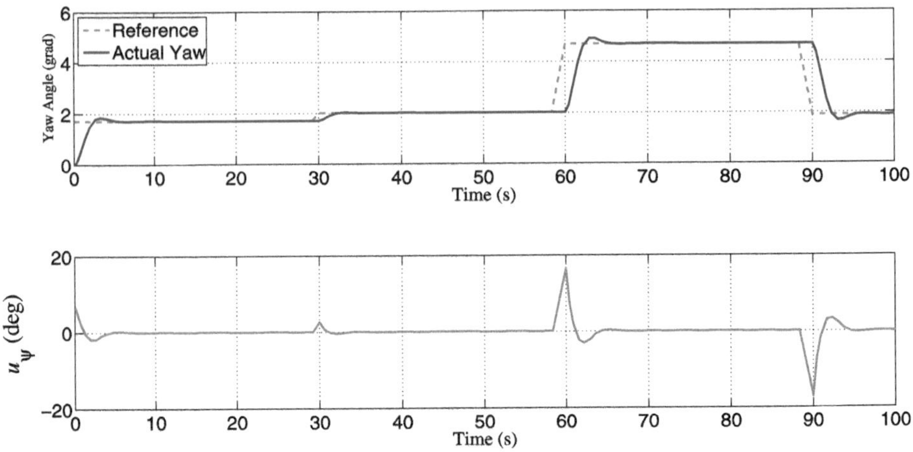

Figure 4.3 Action of the PD controller on yaw.

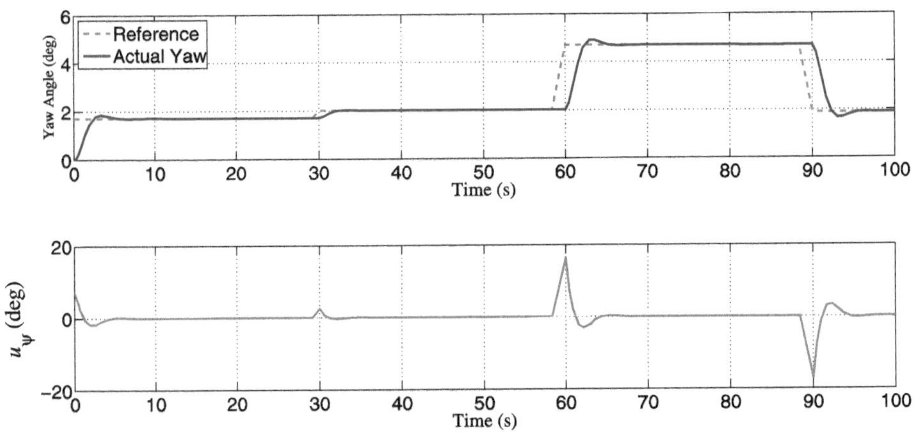

Figure 4.4 Action of the PID controller on yaw.

controller. It is appreciated that the fixed-wing UAV achieved the desired roll angle within 10 seconds. The lower graph of Fig. 4.6 shows the control signal of the PID for the roll angle.

4.2 LQR controller

For the design of the optimal LQR controller we consider Eqs. (3.60)–(3.61) of the dynamic model for altitude, Eqs. (3.63)–(3.64) of the dynamic model for the yaw angle, and Eqs. (3.65)–(3.66) of the dynamic model for the roll angle. These equations

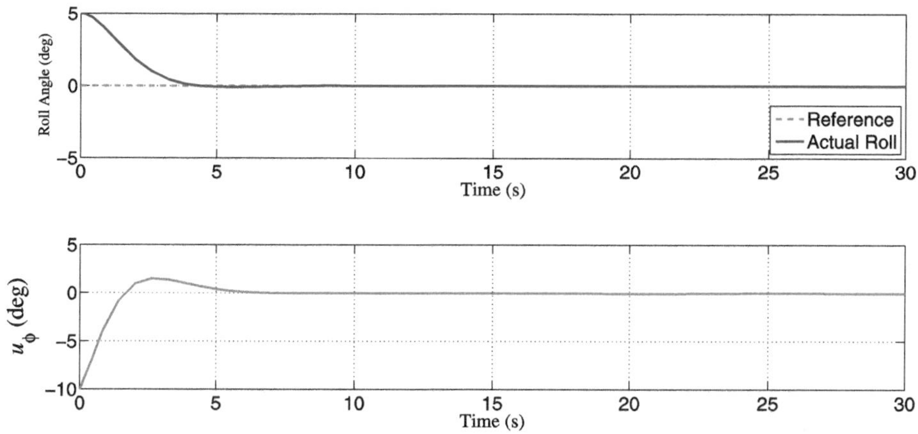

Figure 4.5 Action of the PD controller on roll.

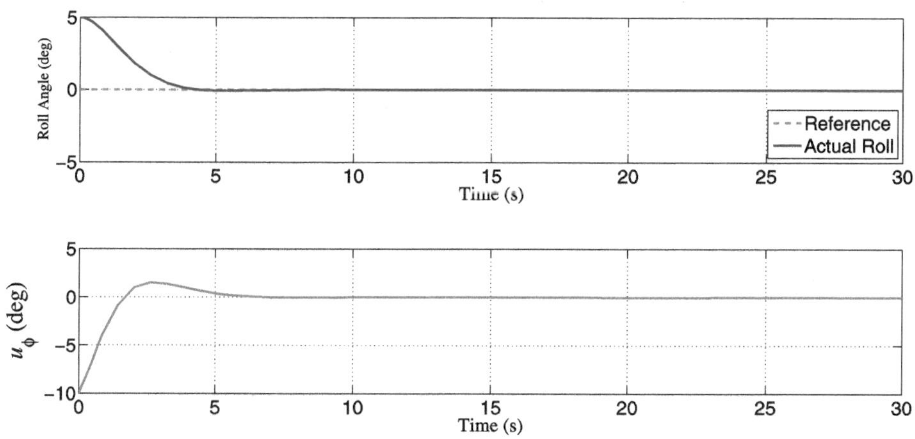

Figure 4.6 Action of the PID controller on roll.

are described in the following form [29]:

$$\dot{\mathbf{x}} = \mathbf{A}\mathbf{x} + \mathbf{B}\mathbf{u},$$

where

$$\mathbf{A} = \begin{pmatrix} 0 & 1 \\ 0 & C_{1l} \end{pmatrix}, \quad \mathbf{B} = \begin{pmatrix} 0 \\ C_{2l} \end{pmatrix},$$

where the subindex l is used to define for which dynamic the LQR controller is designed, thus $l \in \{\theta, \psi, \phi\}$. We aim to determine the matrix $\mathbf{K}$ of the optimal control vector,

$$\mathbf{u}(t) = -\mathbf{K}\mathbf{x}(t), \tag{4.3}$$

which minimizes the cost function

$$J = \int_0^\infty (\mathbf{x}^T \mathbf{Q} \mathbf{x} + \mathbf{u}^T \mathbf{R} \mathbf{u}) dt. \tag{4.4}$$

Note that the second term on the right-hand side of Eq. (4.4) is considered the cost of energy of the control signals. The matrices which determine the relative importance of the error and the cost of energy are **Q** and **R**, where **Q** is a positive definite Hermitian matrix (or positive semidefinite) and **R** is a positive definite Hermitian matrix; a Hermitian matrix is a complex square matrix that is equal to its conjugate transpose.

In the case of the control of the fixed-wing UAV, which is studied in this book, it is supposed that the control vector **u**(t) is not restricted. The lineal control law obtained by Eq. (4.3) is the optimal control law. Therefore, if the unknown elements of the matrix **K** to minimize the cost function J in (4.4) are determined, then $\mathbf{u}(t) = -\mathbf{K}\mathbf{x}(t)$ is optimal for any initial state $\mathbf{x}(0)$. The matrices **Q** and **R** are given by

$$\mathbf{Q} = \begin{pmatrix} 1 & 0 \\ 0 & 1 \end{pmatrix},$$

$$R = \begin{pmatrix} 1 \end{pmatrix}.$$

Assume that u_l is used as control:

$$u_l = -\mathbf{K_l}\mathbf{x_l}.$$

Then, as mentioned for this control law, the matrix of optimal feedback gains $\mathbf{K_l}$ is determined. We can obtain the optimal matrix $\mathbf{K_l}$ by solving the Riccati equation for a positive definite matrix **P**:

$$\mathbf{A}^T\mathbf{P} + \mathbf{P}\mathbf{A} - \mathbf{P}\mathbf{B}\mathbf{R}^{-1}\mathbf{B}^T\mathbf{P} + \mathbf{Q} = 0.$$

The matrix **P** is obtained with the software MATLAB®. This program has a series of programming functions to obtain an LQR controller. The matrix **P** for the altitude dynamics is given by

$$\mathbf{P} = \begin{pmatrix} 1.5640 & 0.3701 \\ 0.3701 & 0.2679 \end{pmatrix}.$$

We substituting this matrix **P** in the following equation of the matrix $\mathbf{K_l}$:

$$\begin{aligned} \mathbf{K_l} &= R^{-1}\mathbf{B}^T\mathbf{P} \\ &= \begin{pmatrix} 1 \end{pmatrix} \begin{pmatrix} 0 & C_{2\theta} \end{pmatrix} \begin{pmatrix} 1.5640 & 0.3701 \\ 0.3701 & 0.2679 \end{pmatrix} = \begin{pmatrix} -1 & -0.7238 \end{pmatrix}, \end{aligned}$$

where $C_{1\theta} = -2.27$ [1/s²] and $C_{2\theta} = -2.702$ [1/s]. Therefore, the LQR controller for the altitude is defined by

$$u_l = -\mathbf{K}_\theta \mathbf{x_l} = -k_{1\theta}(x_{1\theta} - x_{1\theta d}) - k_{2\theta}(x_{2\theta} - x_{2\theta d}), \tag{4.5}$$

where $k_{1\theta}$ and $k_{2\theta}$ are the optimal gains of the LQR controller, $x_{1\theta d}$ represents the pitch angle, and $x_{2\theta d}$ is the component of the angular pitch velocity. For the yaw angle, the matrix **P** obtained with MATLAB is defined by

$$\mathbf{P} = \begin{pmatrix} 2.8231 & 3.3344 \\ 3.3344 & 7.5844 \end{pmatrix}.$$

We substituting this matrix **P** in the following matrix equation $\mathbf{K_l}$:

$$\begin{aligned}\mathbf{K_l} &= R^{-1}\mathbf{B}^T\mathbf{P} \\ &= \begin{pmatrix} 1 \end{pmatrix}\begin{pmatrix} 0 & C_{2\psi} \end{pmatrix}\begin{pmatrix} 2.8231 & 3.3344 \\ 3.3344 & 7.5844 \end{pmatrix} = \begin{pmatrix} 1 & 2.2746 \end{pmatrix},\end{aligned}$$

where $C_{1\psi} = -0.1645\ [1/\mathrm{s}^2]$ and $C_{2\psi} = 0.2999\ [1/\mathrm{s}]$. Therefore, the LQR controller for the yaw angle is defined by

$$u_l = -\mathbf{K}_\psi \mathbf{x_l} = -k_{1\psi}(x_{1\psi} - x_{1\psi d}) - k_{2\psi}(x_{2\psi} - x_{2\psi d}), \tag{4.6}$$

where $k_{1\psi}$ and $k_{2\psi}$ are the optimal gains of the LQR controller, $x_{1\psi d}$ represents the yaw angle, and $x_{2\psi d}$ is the component of the angular velocity of the yaw angle. For the design of the LQR controller for the roll angle, the matrix **P** obtained with MATLAB is given by

$$\mathbf{P} = \begin{pmatrix} 2.6479 & 2.6674 \\ 2.6674 & 4.8688 \end{pmatrix}.$$

We substituting this matrix **P** in the following matrix equation $\mathbf{K_l}$:

$$\begin{aligned}\mathbf{K_l} &= R^{-1}\mathbf{B}^T\mathbf{P} \\ &= \begin{pmatrix} 1 \end{pmatrix}\begin{pmatrix} 0 & C_{2\phi} \end{pmatrix}\begin{pmatrix} 2.6479 & 2.6674 \\ 2.6674 & 4.8688 \end{pmatrix} = \begin{pmatrix} 1 & 1.8253 \end{pmatrix},\end{aligned}$$

where $C_{1\phi} = -0.3084\ [1/\mathrm{s}^2]$ and $C_{2\phi} = 0.3749\ [1/\mathrm{s}]$. Thus, the LQR controller for the roll angle is defined by

$$u_l = -\mathbf{K}_\phi \mathbf{x_l} = -k_{1\phi}(x_{1\phi} - x_{1\phi d}) - k_{2\phi}(x_{2\phi} - x_{2\phi d}), \tag{4.7}$$

where $k_{1\phi}$ and $k_{2\phi}$ are the optimal gains for the LQR controller, $x_{1\phi d}$ represents the roll angle, and $x_{2\phi d}$ is the component of the angular velocity of the roll angle.

In Fig. 4.7 the results obtained with the LQR controller in altitude are shown; it is observed that the fixed-wing UAV achieved the desired altitude in an underdamped form, and stabilizes before 30 seconds; see the upper graph of Fig. 4.7. In the lower graph of Fig. 4.7, the LQR control signal to achieve the desired reference is observed.

Fig. 4.8 shows the action of the LQR controller on the yaw angle. It is observed that in contrast to the PD and PID controllers, the LQR controller did not present the small overshoot in the yaw angle; see the upper graph of Fig. 4.8. Also, it is observed that it applies a small control action in comparison with the PD and PID controllers; see the lower graph of Fig. 4.8.

For the roll angle with the LQR controller, the response is a little underdamped to the desired reference of 0 degrees; see the upper graph of Fig. 4.9. The LQR controller presents small control action on roll in comparison with the PD and PID controllers; see the lower graph of Fig. 4.9.

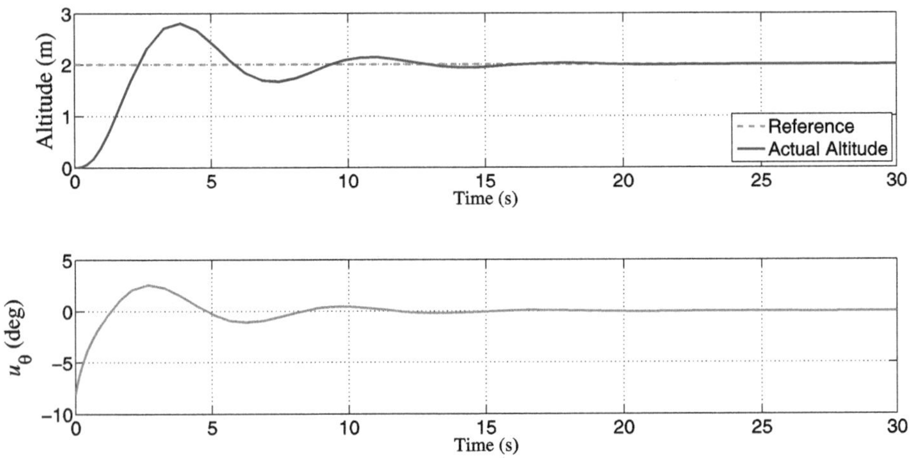

Figure 4.7 Action of the LQR controller on altitude.

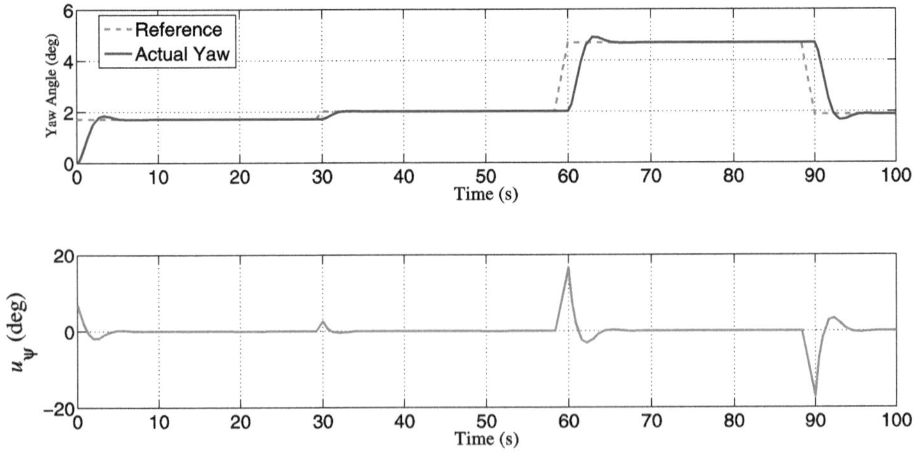

Figure 4.8 Action of the LQR controller on yaw.

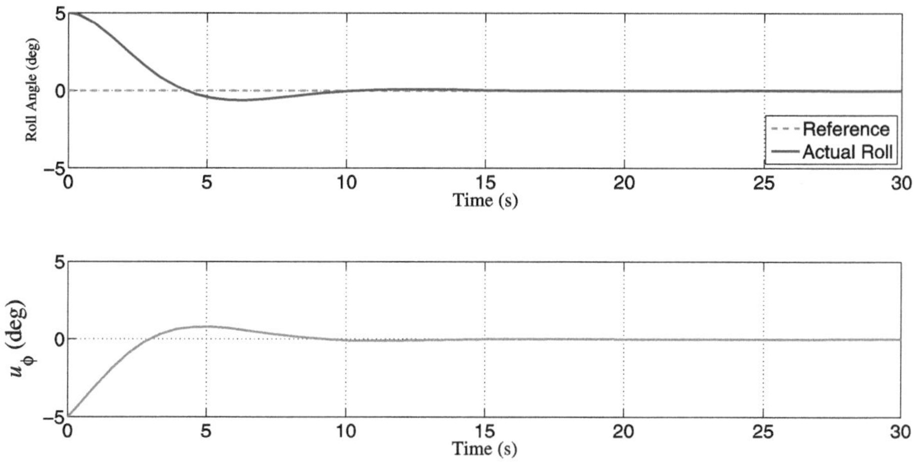

Figure 4.9 Action of the LQR controller on roll.

4.3 LQR controller with the discrete-time Kalman filter

For the development of the LQR controller with the discrete-time Kalman filter [29], we have taken the LQR control structure but added feedback of the angular rates obtained from a discrete-time Kalman filter. We have decided to use such a filter to validate the simple LQR and if possible the solution in the case of IMU failure.

We have two distinct sets of equations that define the discrete-time Kalman filter: time update (prediction) and measurement update (correction). The prediction equations are given as follows [27]. For the development of the LQR controller with the discrete-time Kalman filter [29],

$$\hat{x}_k^- = A\hat{x}_{k-1} + Bu_k, \tag{4.8}$$

$$P_k^- = AP_{k-1}A^T + Q, \tag{4.9}$$

where $\hat{x}_k^-$ is the prior estimate, A_k is the system transition matrix, B_k is the control distribution matrix, u_k is the input vector, P_k^- is the prior error covariance, and Q_k is the covariance noise from the process. The correction equations are given by

$$K_k = P_k^- H^T (HP_k^- H^T + R)^{-1}, \tag{4.10}$$

$$\hat{x}_k = \hat{x}_k^- + K_k(z_k - Hx_k^-), \tag{4.11}$$

$$P_k = (1 - K_k H)P_k^-, \tag{4.12}$$

where K_k is the discrete-time Kalman filter gain, H denotes the measurement matrix, R_k represents the covariance noise from the measurement, $\hat{x}_k$ is the current estimation, z_k is the measurement value, and P_k denotes the error covariance [27]. The parameters used for the discrete-time Kalman filter are shown in Table 4.3.

Table 4.3 Values of the parameters of the discrete-time Kalman filter for the altitude and for the yaw and roll angles.

Parameter	Value
Q	0.01
R	0.01
P	0.0001
H	1

For the altitude control, the discrete-time Kalman filter is applied for the obtention of the pitch angle, thus with Eq. (4.12), it is estimated that $\hat{x}_k = \hat{\theta} = \hat{x}_{1\theta}$. The control law LQR with the discrete-time Kalman filter for altitude is given by

$$u_\theta = -k_{1\theta}(\hat{x}_{1\theta} - x_{1\theta d}) - k_{2\theta}(x_{2\theta} - x_{2\theta d}), \tag{4.13}$$

where $k_{1\theta}$ and $k_{2\theta}$ are the optimal strictly positive gains of the LQR controller with the discrete-time Kalman filter, x_{1ld} represents the desired pitch angle, yaw, or roll, and x_{2ld} is the component of the desired angular velocity (pitch, yaw, or roll).

The results obtained in altitude with the LQR controller with discrete-time Kalman filter are shown in Fig. 4.10. It is observed that the fixed-wing UAV achieved the desired altitude in an underdamped form, which stabilizes within 20 seconds; see the upper graph of Fig. 4.10. In the lower graph of Fig. 4.10, the LQR control signal with discrete-time Kalman filter to achieve such a reference is observed.

For the LQR controller with the discrete-time Kalman filter in yaw the discrete-time Kalman filter has been applied to estimate the component of the angular velocity in the direction of the yaw angle $\hat{x}_k = \hat{r} = \hat{x}_{2\psi}$. This controller is given by

$$u_\psi = -k_{1\psi}(x_{1\psi} - x_{1\psi d}) - k_{2\psi}(\hat{x}_{2\psi} - x_{2\psi d}), \tag{4.14}$$

where $k_{1\psi}$ and $k_{2\psi}$ are the optimal gains of the LQR controller with the discrete-time Kalman filter in yaw, and the simulation results obtained are shown in Fig. 4.11. It is observed that the fixed-wing UAV achieved the desired values for the yaw angle in an underdamped form, and it stabilizes within 20 seconds; see the upper graph of Fig. 4.11. In the lower graph of Fig. 4.11, the LQR control signal with the discrete-time Kalman filter to achieve the references of the yaw angle is presented.

For the LQR controller with the discrete-time Kalman filter in roll, the filter mentioned for the estimate of the angular velocity in roll $\hat{x}_k = \hat{p} = \hat{x}_{2\phi}$ using the same equation (4.11) with the dynamic of the roll is used. Thus, the LQR controller with discrete-time Kalman filter is given by

$$u_\phi = -k_{1\phi}(x_{1\phi} - x_{1\phi d}) - k_{2\phi}(\hat{x}_{2\phi} - x_{2\phi d}), \tag{4.15}$$

where $k_{1\phi}$ and $k_{2\phi}$ are the optimal gains of the LQR controller with discrete-time Kalman filter in the roll. The results obtained with this controller for the roll are appre-

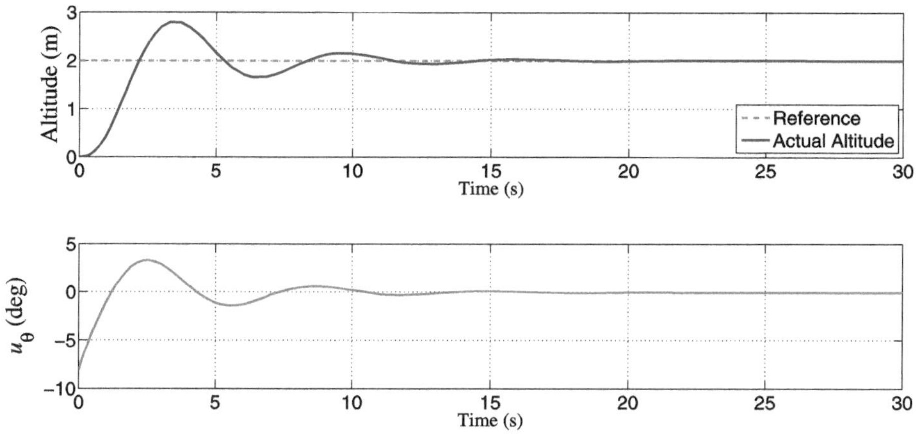

Figure 4.10 Action of the LQR controller with the discrete-time Kalman filter on altitude.

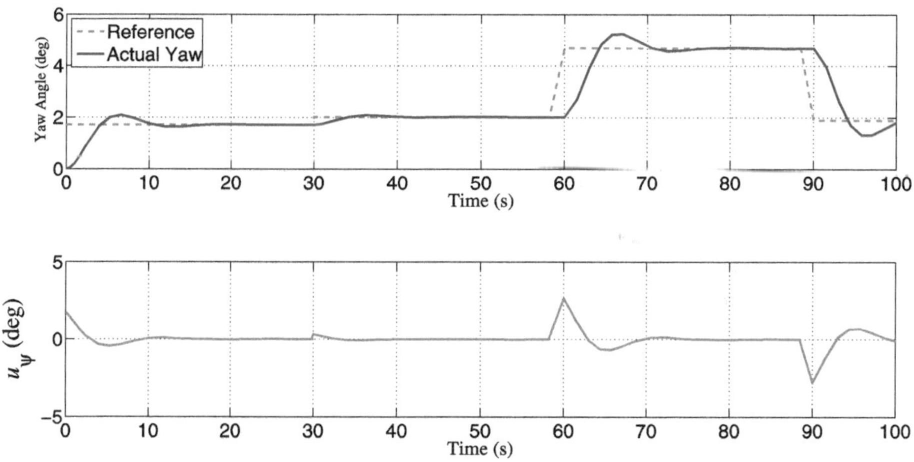

Figure 4.11 Action of the LQR controller with the discrete-time Kalman filter on yaw.

ciated in Fig. 4.12. It is observed that the fixed-wing UAV converges to the desired roll angle in an underdamped form and stabilizes within 25 seconds; see the upper graph of Fig. 4.12. In the lower graph of Fig. 4.12, the LQR control signal with discrete-time Kalman filter to achieve the roll angle reference is presented.

The optimal values of the gains of the LQR controller and the LQR controller with the discrete-time Kalman filter are shown in Table 4.4.

To make a comparison between controllers the $\mathcal{L}_2$ norm [41] is used; this norm is applied to all results obtained in the simulations for the linear controllers and nonlinear controllers and combinations of nonlinear controllers. Then, the $\mathcal{L}_2$ norm of the error and the actions of each controller studied in this book are obtained. Thus, the $\mathcal{L}_2$

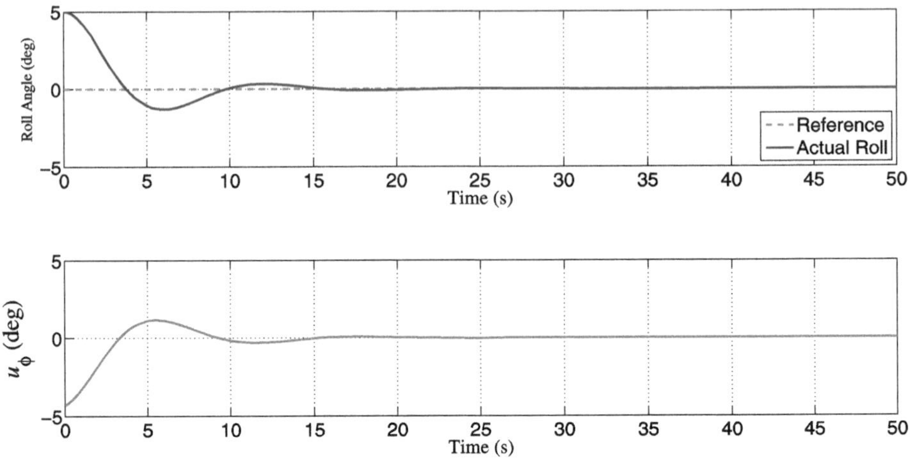

Figure 4.12 Action of the LQR controller with the discrete-time Kalman filter on roll.

Table 4.4 Gains of the LQR controller and the LQR controller with the discrete-time Kalman filter.

Movement	k_{1I}	k_{2I} [s]
Altitude	1	0.7238
Yaw	1	2.2746
Roll	1	1.8253

norm for the error is given by

$$\mathcal{L}_2[e_I] = \sqrt{\frac{1}{T-t_0}\int_{t_0}^{T} \|e_I\|^2 dt}. \tag{4.16}$$

The $\mathcal{L}_2$ norm to obtain the action of the control law is defined as

$$\mathcal{L}_2[u_I] = \sqrt{\frac{1}{T-t_0}\int_{t_0}^{T} \|u_I\|^2 dt}. \tag{4.17}$$

In Table 4.5 the results obtained with the $\mathcal{L}_2$ norm are shown. The linear controller with less action on the three analyzed movements (altitude, yaw, and roll) was the LQR controller with the discrete-time Kalman filter. It is also observed that the LQR controller with the discrete-time Kalman filter generated less action compared to the PD, PID, and LQR controllers.

Table 4.5 $\mathcal{L}_2$ norm for the errors and the actions of the linear controllers.

Controller	Altitude [m]		Yaw [deg]		Roll [deg]	
	$\mathcal{L}_2[e_\theta]$	$\mathcal{L}_2[u_\theta]$	$\mathcal{L}_2[e_\psi]$	$\mathcal{L}_2[u_\psi]$	$\mathcal{L}_2[e_\phi]$	$\mathcal{L}_2[u_\phi]$
PD	0.1341	1.1867	0.0810	0.0978	0.0765	0.1347
PID	0.1135	1.2236	0.0810	0.0978	0.0765	0.1345
LQR	0.1053	0.1744	0.0718	0.0194	0.1291	0.0942
LQG	0.0824	0.1176	0.0704	0.0170	0.0744	0.0668

4.4 Concluding remarks

This chapter described the design of linear controllers. For the development of the techniques used, the aerodynamic model that defines the fixed-wing UAV is shown in Chapter 3. Moreover, simulation results obtained with the linear controllers are shown. Besides, an error analysis using the vectorial norms was performed to know what control law presented the more significant error.

The error and action of said controllers to achieve desired values were analyzed for the three fundamental movements of a fixed-wing UAV: altitude, yaw angle, and roll angle.

Nonlinear controllers

5

This chapter presents the design of nonlinear controllers; these controllers, as linear controllers, are applied to the altitude and the yaw and roll angles. The nonlinear controllers designed are as follows: the nested saturation technique is presented in Section 5.1, backstepping control is developed in Section 5.2, and Section 5.3 is devoted to sliding mode control.

Also, considering the nonlinear control techniques mentioned above, such methods are combined to obtain control laws with increased robustness. The resulting controllers are the following: nested saturation with sliding mode is presented in Section 5.4, Section 5.5 describes the nested saturation with second-order sliding mode, and nested saturation with high-order sliding mode is described in Section 5.6. The backstepping technique with sliding mode is presented in Sections 5.7–5.9.

Finally, adaptive control is developed based on the MIT rule with the sliding mode theory. This chapter presents the simulation results obtained with the nonlinear controllers mentioned above.

5.1 Nested saturation controller

In this section, the application of the nested saturation controller to a fixed-wing UAV is described [40], [63]. The nested saturation controller notation uses the change in variables in Section 3.6.3. Then the dynamics are represented by

$$\dot{x}_{1l} = x_{2l}, \tag{5.1}$$

$$\dot{x}_{2l} = C_{1l}x_{2l} + C_{2l}u_l, \tag{5.2}$$

where we will continue to use the subscript l to indicate the movement to which the control law is applied, thus $l \in \{\theta, \psi, \phi\}$. Then, a linear transformation to the system defined by (5.1)–(5.2) is performed, and this yields a linear transformation matrix given by

$$\begin{bmatrix} z_{1l} \\ z_{2l} \end{bmatrix} = \begin{bmatrix} a_{1l}/C_{1l} & 1 \\ 0 & 1 \end{bmatrix} \begin{bmatrix} x_{1l} \\ x_{2l} \end{bmatrix}, \tag{5.3}$$

where a_{1l} is a definite positive constant and C_{1l} is the derivative of aerodynamic stability. For this controller a conventional saturation function $\sigma_b : \mathbb{R} \to \mathbb{R}$ is used, which has a limit b such that

$$\sigma_b(s) = \begin{cases} -b, & s < -b, \\ s, & -b \le s \le b, \\ b, & s > b, \end{cases} \tag{5.4}$$

Design of Control Laws and State Observers for Fixed-Wing UAVs. https://doi.org/10.1016/B978-0-32-395405-1.00014-5

Table 5.1 Gains of the nested saturation controller.

Movement	a_{1l} [1/s^2]	k_{1l} [1/s]
Altitude	8	0.1
Yaw	0.25	1
Roll	0.5	1

where b is a positive constant. Then, for the system given by (5.1)–(5.2), with the linear transformation (5.3) and with the use of the saturation function (5.4), the control by nested saturation is given by

$$u_l = \frac{-C_{1l}z_{2l} - \sigma_{2l}[k_{1l}z_{2l} + \sigma_{1l}(k_{1l}z_{1l})]}{C_{2l}}, \tag{5.5}$$

where k_{1l} is a positive definite gain, $z_{1l} = \frac{a_{1l}}{C_{1l}}\tilde{e}_l + \dot{\tilde{e}}_l$, and $z_{2l} = \dot{\tilde{e}}_l$. The definition of the errors e_l is the same as used for linear controllers. Remember that the error for altitude has been defined as $\tilde{e}_h = h_d - h$, where h_d is the desired altitude and h is the obtained altitude of the integral equation (3.62).

The altitude control is achieved by varying the pitch angle θ. Therefore, for a change of θ, an error of the form $\tilde{e}_\theta = \theta_d - \theta$ is defined, where $\theta_d = \arctan(\tilde{e}_h/\varsigma)$ and θ_d is the desired pitch angle, where ς is the distance from the center of mass of the fixed-wing UAV to the nose of the UAV.

The values of the gains of the nested saturation controller are shown in Table 5.1. The values for the controller gains for altitude were obtained so that the plane's altitude converges at the desired altitude in less than 30 seconds and without overshoot. The control action was not very large. For yaw and roll angles, it is sought that the responses reach the desired reference in less than 10 seconds without overshoot and with a small control action.

For the yaw controller, an error $\tilde{e}_\psi = \psi_d - \psi$ is defined, where ψ_d is the desired yaw angle and ψ is obtained from the integral of Eq. (3.63). Finally, to lead the fixed-wing UAV to the desired roll, the error is denoted as $\tilde{e}_\phi = \phi_d - \phi$, where ϕ_d is the desired roll angle and ϕ is defined as the roll angle obtained by the integral of Eq. (3.65). The results of the nested saturation controller for altitude are shown in Fig. 5.1. In the lower graph of Fig. 5.1, the control action applied by the controller based on nested saturation to achieve the desired altitude is shown. It is observed that the fixed-wing UAV achieves the desired altitude in less than 30 seconds (see the upper graph of Fig. 5.1).

For the simulation of the nested saturation controller for the yaw angle, observe Fig. 5.2, in which a signal is referenced with different magnitudes. These references are the same ones that have been used for the linear controllers used in the previous chapter for this same angle. In the upper graph of Fig. 5.2, the convergence of the yaw angle to the desired reference is shown; it is appreciated that yaw angles are achieved in less than 10 seconds for every reference simulated in Fig. 5.2. Besides, in the lower graph of Fig. 5.2, the control signals to achieve the desired references are presented.

For the simulation of the roll angle, it is considered that the fixed-wing UAV has an initial roll angle of 5 degrees, and the control law should lead the roll angle to the desired reference of zero degrees. Fig. 5.3 shows the results with the nested saturation controller for roll angle. In the upper graph of Fig. 5.3, it is observed how the controller leads to the desired reference of the fixed-wing UAV, that is, 0 degrees, in less than 5 seconds. In the lower graph of Fig. 5.3, the control action that the nested saturation controller applied is shown.

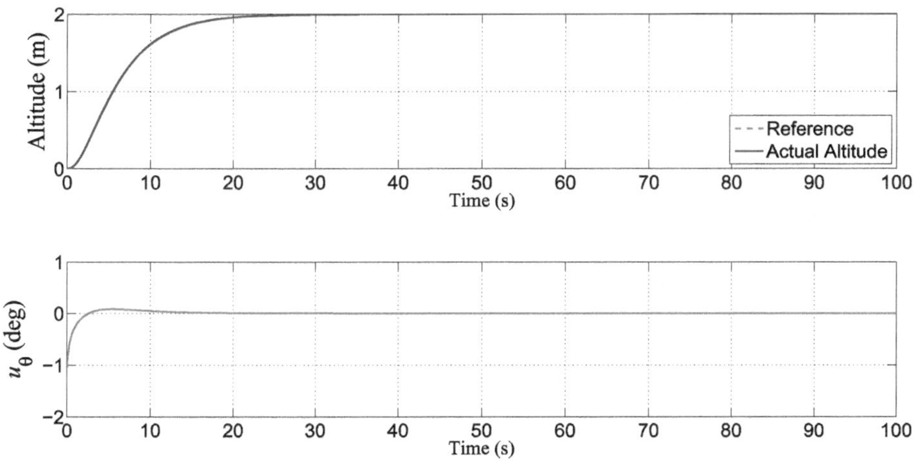

Figure 5.1 Action of the nested saturation controller on altitude.

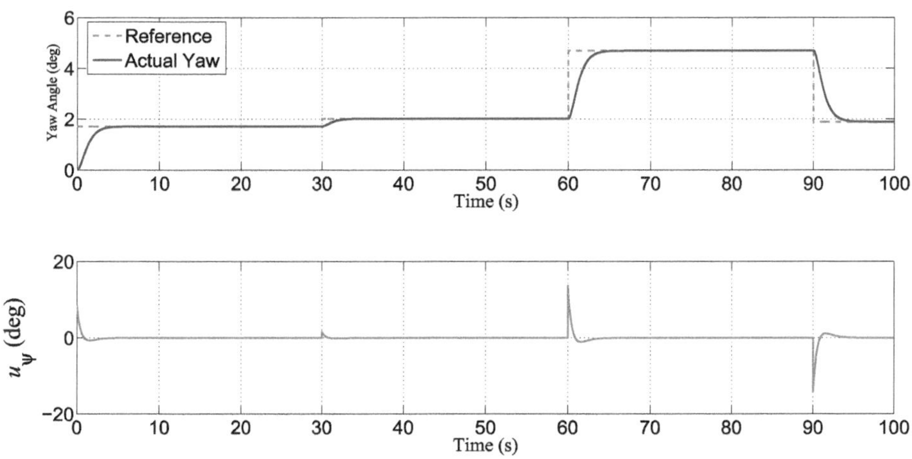

Figure 5.2 Action of the nested saturation controller on yaw.

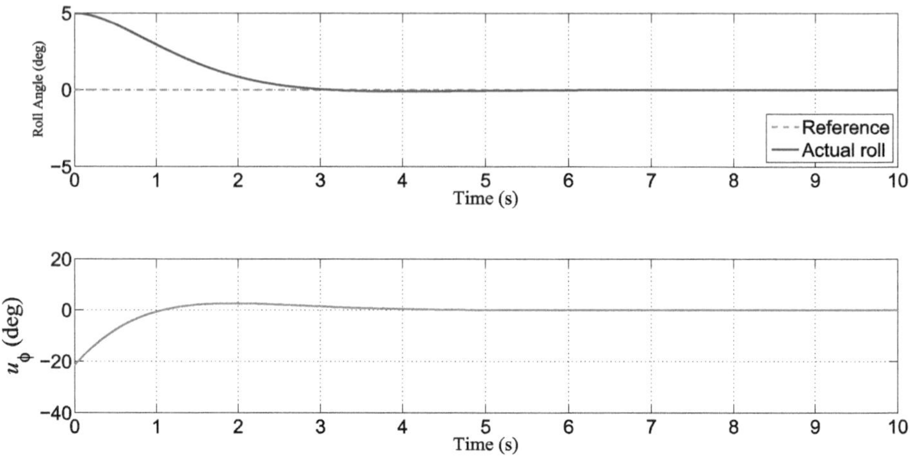

Figure 5.3 Action of the nested saturation controller on roll.

5.2 Backstepping controller

The backstepping controller has a defined model, as the method of nested saturation control [41]:

$$\dot{x}_{1l} = x_{2l}, \tag{5.6}$$

$$\dot{x}_{2l} = C_{1l}x_{2l} + C_{2l}u_l. \tag{5.7}$$

The first step to designing the backstepping controller is to define the error as

$$\tilde{e}_l = x_{1ld} - x_{1l}, \tag{5.8}$$

where x_{1ld} is the desired angular position, x_{1l} is the angular position, and l is the angle of interest for the controller $l = \theta, \psi, \phi$. It is defined a Lyapunov candidate function for the error:

$$V_1 = \frac{1}{2}\tilde{e}_l^2. \tag{5.9}$$

Taking the derivative of (5.9),

$$\dot{V}_1 = \tilde{e}_l\dot{\tilde{e}}_l \tag{5.10}$$

$$= \tilde{e}_l(\dot{x}_{1ld} - \dot{x}_{1l}) \tag{5.11}$$

$$= \tilde{e}_l(\dot{x}_{1ld} - x_{2l}). \tag{5.12}$$

Now x_{2l} is taken as a virtual control input:

$$x_{2l} = \dot{x}_{1ld} + k_{1l}\tilde{e}_l - z_{1l}, \tag{5.13}$$

Table 5.2 Gains of the backstepping controller.

Movement	k_{1l} [1/s]	k_{2l} [1/s]	k_{3l} [1/s^2]
Altitude	3	6	0.9
Yaw	1.5	1.5	0.9
Roll	1.5	1.5	0.2

where z_{1l} is a new variable of state introduced to apply this control technique and k_{1l} is a definite positive gain. Now, substituting x_{2l} in (5.12) we obtain

$$\dot{V}_1 = z_{1l}\tilde{e}_l - k_{1l}\tilde{e}_l^2. \tag{5.14}$$

Now, let us consider a Lyapunov candidate function:

$$V_2 = V_1 + \frac{1}{2}z_{1l}^2, \tag{5.15}$$

where V_1 is defined in (5.9). Taking the derivative of (5.15),

$$\dot{V}_2 = \dot{V}_1 + z_{1l}\dot{z}_{1l} \tag{5.16}$$
$$= k_{3l}(z_{1l}\tilde{e}_l - k_{1l}\tilde{e}_l^2) + z_{1l}(\ddot{x}_{1ld} + k_{1l}\dot{\tilde{e}}_l - \dot{x}_{2l}) \tag{5.17}$$
$$= k_{3l}(z_{1l}\tilde{e}_l - k_{1l}\tilde{e}_l^2) + z_{1l}(\ddot{x}_{1ld} + k_{1l}\dot{\tilde{e}}_l - C_{1l}x_{2l} - C_{2l}u_l). \tag{5.18}$$

Given that the considered dynamic model contains only two states and that the control input u_l appears in (5.18), the control law backstepping can be proposed as

$$u_l = \frac{C_{1l}x_{2l} - \ddot{x}_{1ld} - k_{1l}\dot{\tilde{e}}_l - k_{2l}z_{1l} - k_{3l}\tilde{e}_l}{-C_{2l}}, \tag{5.19}$$

where k_{1l}, k_{2l}, and k_{3l} are strictly positive gains. Finally, the derivative of the Lyapunov candidate function (5.15) is

$$\dot{V}_2 = -k_{1l}k_{3l}\tilde{e}_l^2 - k_{2l}z_{1l}^2 \leq 0. \tag{5.20}$$

The gains of the backstepping controller are presented in Table 5.2. The gains were tuned in a heuristic form, so there was no overshoot to converge to the desired reference and the reference was reached in less than 10 seconds.

Fig. 5.4 shows the results obtained with the backstepping controller for altitude. In the upper graph of Fig. 5.4, it is observed that the fixed-wing UAV reaches the desired reference in less than 10 seconds. The lower graph of Fig. 5.4 presents the necessary control action to reach the control objective.

The yaw angle control is used for the same signal with different values of reference. It is observed that the objective of control in the yaw angle is achieved in much less than 10 seconds (see the upper graph of Fig. 5.5). In the lower graph of the same figure the necessary control action is shown.

Fig. 5.6 shows the results obtained with the backstepping controller for roll. It is appreciated that the roll angle converges to the desired reference in less than 5 seconds

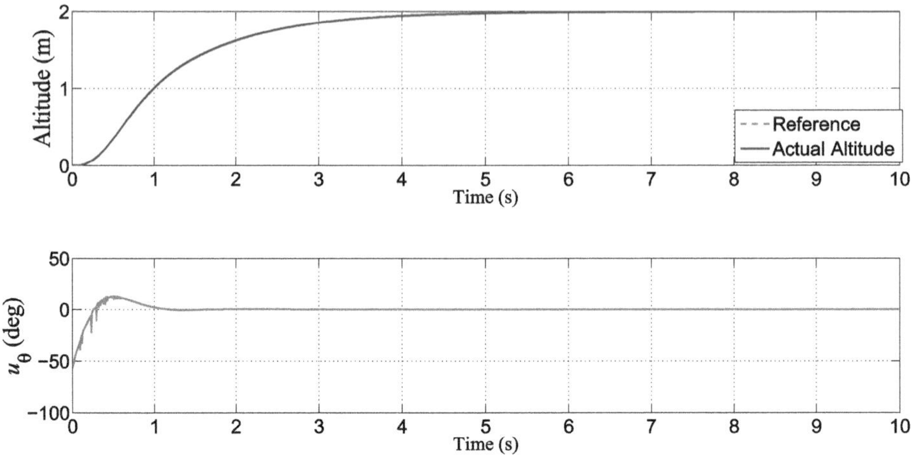

Figure 5.4 Action of the backstepping controller on altitude.

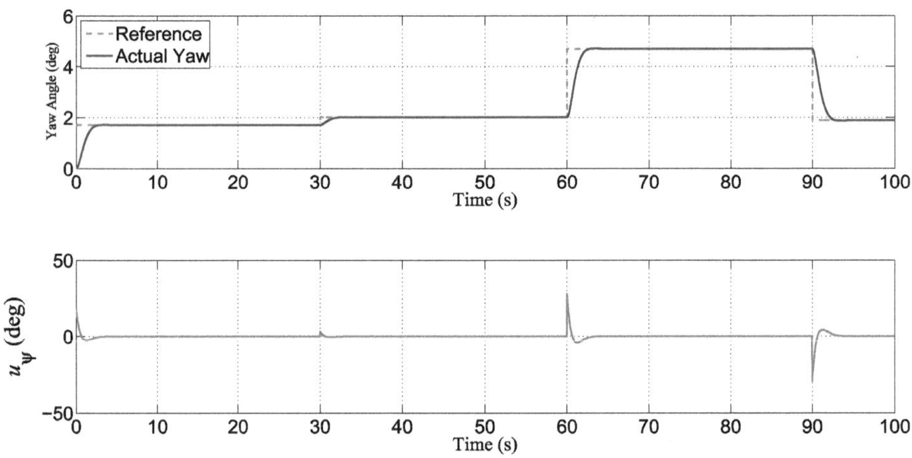

Figure 5.5 Action of the backstepping controller on yaw.

(see the upper graph of Fig. 5.6). In the lower graph of Fig. 5.6 the control action to achieve this response is shown.

5.3 Sliding mode controller

Consider the following second-order system again [41]:

$$\dot{x}_{1l} = x_{2l}, \tag{5.21}$$

$$\dot{x}_{2l} = C_{1l}x_{2l} + C_{2l}u_l, \tag{5.22}$$

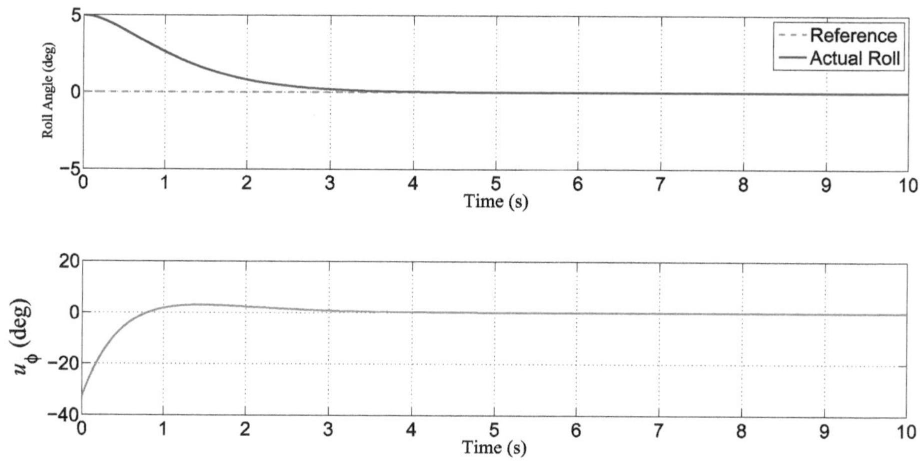

Figure 5.6 Action of the backstepping controller on roll.

where C_{1l} and C_{2l} are the aerodynamic stability derivatives. Now the technique of sliding mode control is applied. The sliding manifold is defined as

$$s = \dot{\tilde{e}}_l + k_{1l}\tilde{e}_l = 0, \tag{5.23}$$

where k_l is a positive definite gain, $\tilde{e}_l$ is the error of the angular position, and l represents either the pitch angle θ, yaw ψ, or roll ϕ $l \in \{\theta, \psi, \phi\}$. Taking the derivative of the sliding manifold (5.23) yields

$$\dot{s} = \dot{x}_{2ld} - \dot{x}_{2l} + k_{1l}(\dot{x}_{1ld} - x_{2l}) \tag{5.24}$$
$$= \dot{x}_{2ld} + C_{1l}x_{2l} + C_{2l}u_l - k_{1l}(\dot{x}_{1ld} - x_{2l}). \tag{5.25}$$

It is observed that in (5.25) the control input u_l appears. Then, to eliminate the terms of the right-hand side of (5.25) a sliding mode controller is designed with the following structure:

$$u_l = \frac{C_{1l}x_{2l} - \dot{x}_{2ld} - k_{1l}(\dot{x}_{1ld} - x_{2l}) - \beta_{xl}\,\mathrm{sgn}(s)}{C_{2l}}, \tag{5.26}$$

where the sign function is defined by

$$\mathrm{sgn}(s) = \begin{cases} 1, & s > 0, \\ 0, & s = 0, \\ -1, & s < 0 \end{cases}$$

and β_{xl} is a strictly positive gain. The gain values used for the simulations of the sliding mode controller are shown in Table 5.3. The gains were tuned in a heuristic form, but aiming not to have overshoot when the reference is reached, with convergence toward the desired reference in time than 10 seconds, and with a small control action.

Table 5.3 Gains of the sliding mode controller.

Movement	k_{1l} [1/s]	β_{x1l} [1/s]
Altitude	2	0.1
Yaw	1	0.1
Roll	1	0.1

In Fig. 5.7 results obtained with the sliding mode controller are presented. In the upper graph of Fig. 5.7, it is observed how the altitude converges to the desired reference within 10 seconds. In the lower graph of Fig. 5.7, the control action generated by the controller is shown; the chattering effect produced by this control technique is appreciated.

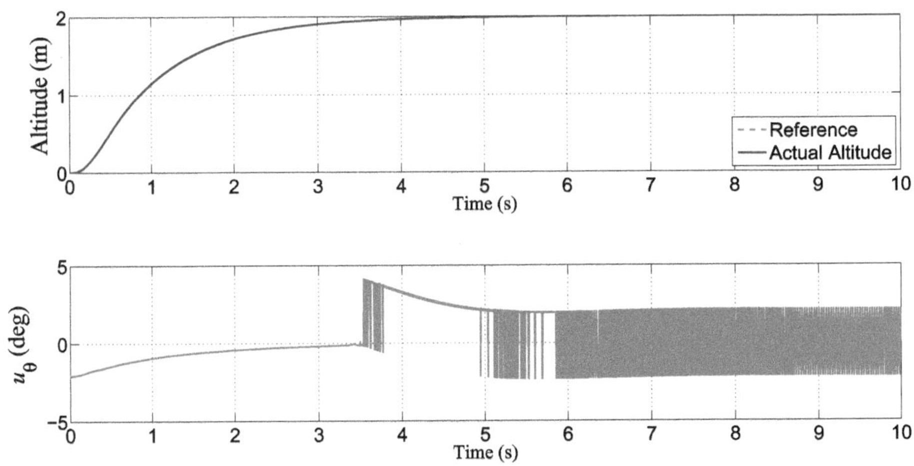

Figure 5.7 Action of the sliding mode controller on altitude.

Fig. 5.8 shows the results obtained with sliding mode control for yaw. In the upper graph of the figure, the convergence of the yaw angle to the desired reference is observed; for this simulation, different references were used again, and it is appreciated that the convergence is achieved in less than 10 seconds. The lower graph of Fig. 5.8 shows the control action generated by the controller and also presents the chattering effect.

In Fig. 5.9 the simulation results obtained for the roll angle are presented. For this simulation, the roll angle has an initial angle of 10 degrees and the sliding mode controller leads the roll angle to 0 degrees. It is observed in the upper graph of this figure that the control objective is achieved within 8 seconds. The control action with the chattering effect in the controller's output is shown in the lower graph of the same figure.

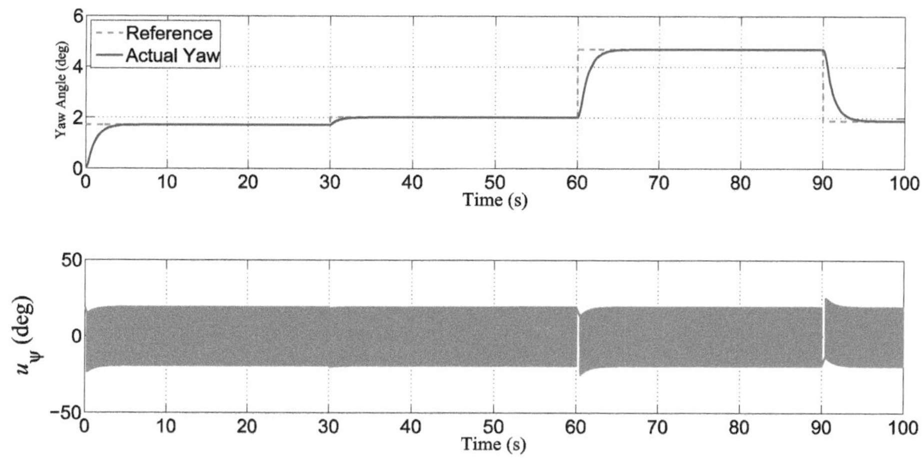

Figure 5.8 Action of the sliding mode controller on yaw.

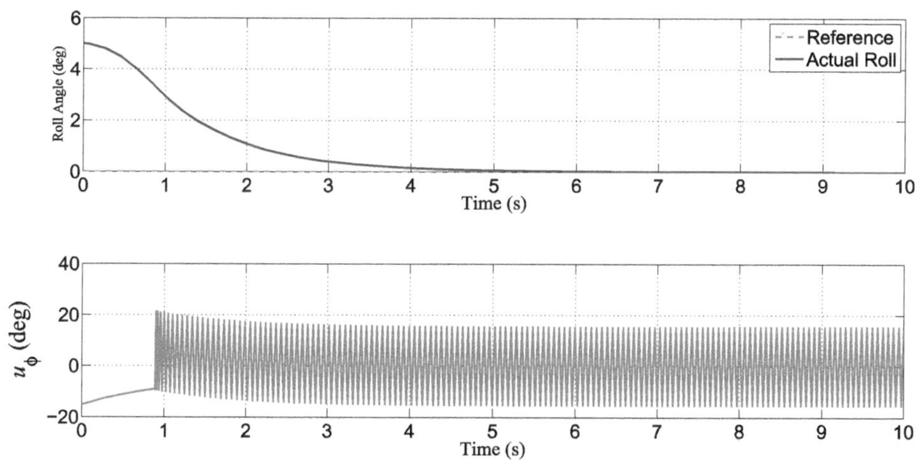

Figure 5.9 Action of the sliding mode controller on roll.

5.4 Nested saturation with sliding mode

The nested saturation controller with sliding mode has been designed by the union or combination of the controllers described in Sections 5.1 and 5.3. We begin from the uncoupled model:

$$\dot{x}_{1l} = x_{2l}, \tag{5.27}$$

$$\dot{x}_{2l} = C_{1l}x_{2l} + C_{2l}u_l. \tag{5.28}$$

Let us consider the control law (5.5) and a sliding manifold $s = \dot{\tilde{e}}_l + k_{2l}\tilde{e}_l = 0$, where k_{2l} is a strictly positive gain and l is as before, that is, $l \in \{\theta, \psi, \phi\}$, this describes the

Table 5.4 Gains of the nested saturation controller with sliding mode.

Movement	a_{1l} [1/s^2]	k_{1l} [1/s]	k_{2l} [1/s]	β_{x1l} [1/s]
Altitude	4	0.9	0.1	0.18
Yaw	0.2	0.8125	1	0.05
Roll	0.5	0.8125	1.25	0.05

angle for the designed controller. The nested saturation controller with sliding mode can be defined as

$$u_l = \frac{-C_{1l}z_{2l} - \beta_{x1l}\,\mathrm{sgn}(s) - \sigma_{2l}[k_{1l}z_{2l} + \sigma_{1l}(k_{1l}z_{1l})]}{C_{2l}}, \tag{5.29}$$

where β_{x1l}, k_{1l}, and k_{2l} are strictly positive constants and z_{1l} and z_{2l} are defined in Section 5.1 for the design of the nested saturation controller. Table 5.4 shows the values of the gains for the simulation of the nested saturation controller with sliding mode. The tuning of this controller is realized with the same considerations as for the nested saturation controller with sliding mode mentioned before.

Fig. 5.10 presents the response obtained with the nested saturation controller with sliding mode. We can observe in the upper graph that the altitude of the fixed-wing UAV achieved the reference or the desired altitude in less than 10 seconds. The lower graph shows the applied control action, which presents the chattering effect.

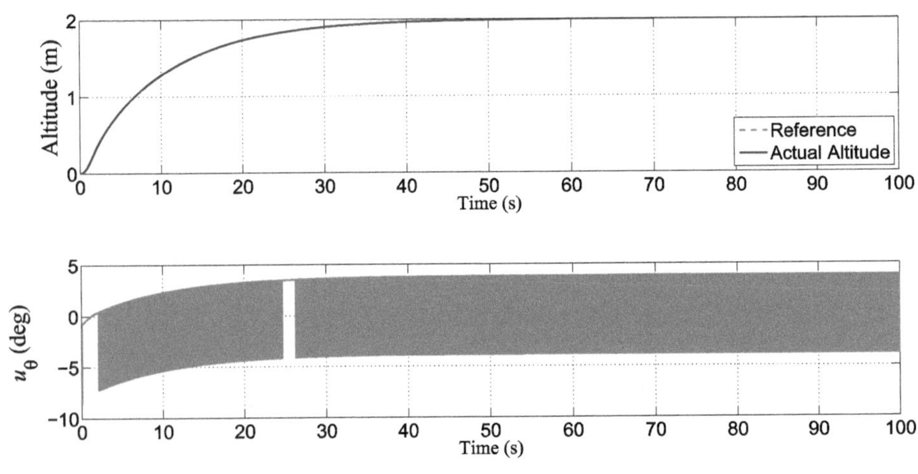

Figure 5.10 Action of the nested saturation controller with sliding mode on altitude.

For the nested saturation controller with sliding mode for the yaw angle, the upper graph of Fig. 5.11 shows the yaw angle, reaching the different reference values, and in the lower graph, we can appreciate the control action supplied by the controller;

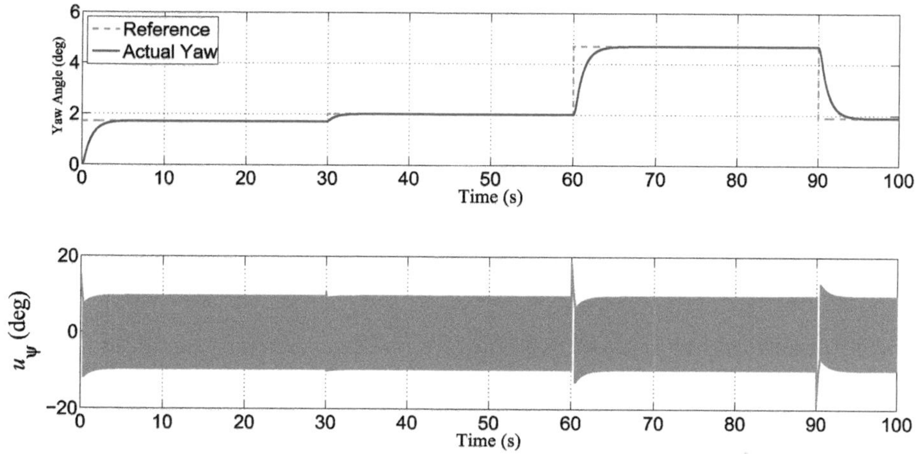

Figure 5.11 Action of the nested saturation controller with sliding mode on yaw.

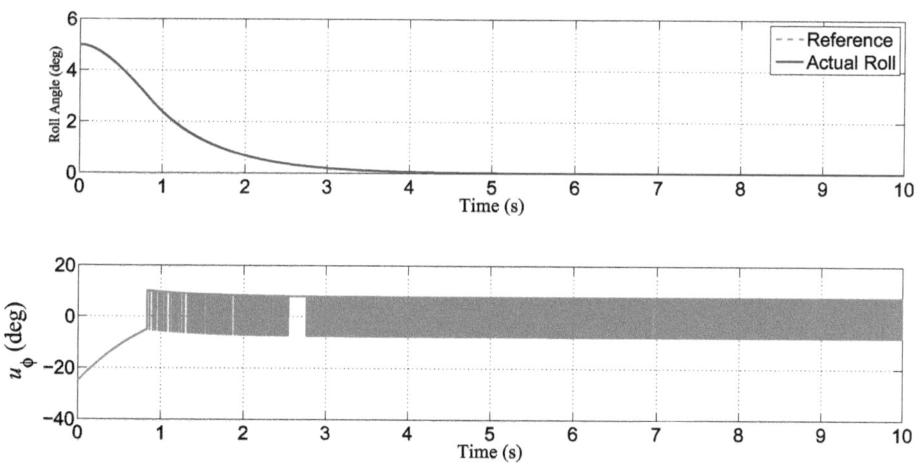

Figure 5.12 Action of the nested saturation controller with sliding mode on roll.

it is worth mentioning which this controller exhibits a chattering effect in its control output.

For the roll angle, it is considered that the fixed-wing UAV has an initial roll angle of 5 degrees, and the controller based on nested saturation with sliding mode is used to lead to a roll angle of 0 degrees. The upper graph of Fig. 5.12 shows the obtained results, while the lower graph of the same figure shows the control action with the chattering effect.

5.5 Nested saturation with 2-SM

Designing the nested saturation controller with second-order sliding mode (2-SM) is necessary to design a robust first-order differentiator. Because a derivative in real-time is sensitive to noise at the moment of the derivative action [47], this robust differentiator in real-time is given by [48]

$$\dot{x}_0 = v_0 = -\lambda_0 |x_0 - s_l|^{1/2} \operatorname{sgn}(x_0 - s_l) + x_1, \tag{5.30}$$

$$\dot{x}_1 = -\lambda_1 \operatorname{sgn}(x_1 - v_0), \tag{5.31}$$

where x_0 and x_1 are estimates in real-time of s_l and $\dot{s}_l$, respectively, and λ_0 and λ_1 are constants for the tuning of the differentiator; in this case, the used values are $\lambda_0 = 12$ and $\lambda_1 = 6.6$, which were obtained as shown in reference [48]. For this controller was designed using nested saturation and second-order sliding mode. We took as reference the controller obtained in (5.29) and added the derivative of the sliding manifold with its respective sign function to reduce the chattering effect that it has in the control output. Therefore, the nested saturation controller with second-order sliding mode is given as

$$u_l = \frac{-C_{1l} z_{2l} - \beta_{x1l} \operatorname{sgn}(s) - \beta_{x2l} \operatorname{sgn}(\dot{s}) - \sigma_{2l}[k_{1l} z_{2l} + \sigma_{1l}(k_{2l} z_{1l})]}{C_{2l}}, \tag{5.32}$$

where β_{x1l}, β_{x2l}, k_{1l}, and k_{2l} are strictly positive constants and z_{1l} and z_{2l} were defined in the design of the nested saturation controller (5.32). Table 5.5 shows the values of the used gains by the nested saturation with second-order sliding mode This controller was tuned with the same considerations for the nested saturation controller with sliding mode mentioned in the previous sections, but the gains β_{x1l} and β_{x2l} were fine-tuned to reduce the chattering effect.

In Fig. 5.13 the results obtained for the control of altitude with the nested saturation controller with second-order sliding mode are presented. In the upper graph of Fig. 5.13, it is observed that the desired altitude is achieved within 40 seconds, in contrast to the nested saturation controller and the nested saturation controller with sliding mode, which achieve the reference in less than 10 seconds. However, in the lower graph of Fig. 5.13 it is appreciated that the control action caused less chattering than the sliding mode controller and the nested saturation controller with sliding mode.

The results of the nested saturation controller with second-order sliding mode for the yaw angle are shown in Fig. 5.14. In the upper graph, it is shown that the yaw angle achieves the reference values in less than 10 seconds. In the lower graph, the control action is observed with reduced chattering compared to the sliding mode controller and with the nested saturation controller.

The roll angle had an initial angle of 5 degrees and we aimed to bring the roll angle toward 0 degrees. In the upper graph of Fig. 5.15, it is observed that this was accomplished in less than 5 seconds, and in the lower graph the reduction of the chattering effect with this controller is shown.

Table 5.5 Gains of the nested saturation controller with second-order sliding mode.

Movement	a_{1l} [1/s^2]	k_{1l} [1/s]	k_{2l} [1/s]	β_{x1l} [1/s]	β_{x2l} [s^2]
Altitude	4	0.9	0.1	0.1	0.096
Yaw	0.49	0.8125	1.25	0.1	0.096
Roll	0.49	0.8125	1.25	0.1	0.096

Figure 5.13 Action of the nested saturation controller with 2-SM on altitude.

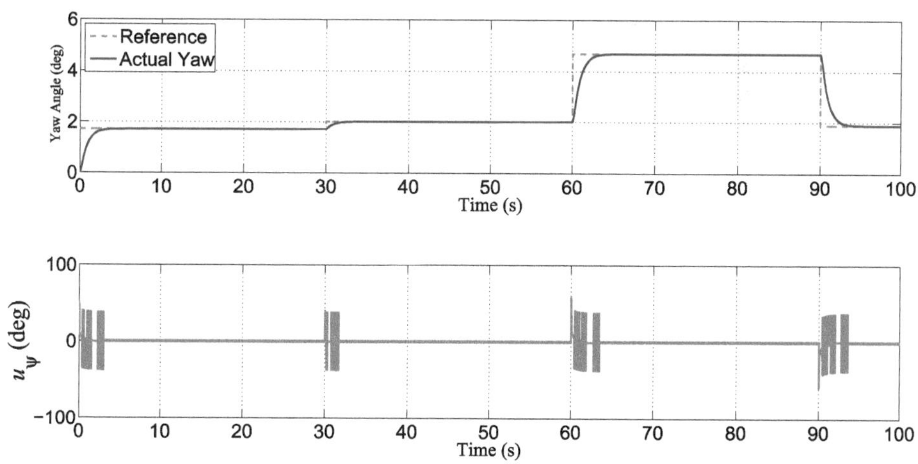

Figure 5.14 Action of the nested saturation controller with 2-SM on yaw.

5.6 Nested saturation with HOSM

A controller with second-order sliding mode can be enough to reduce the chattering effect of sliding mode controllers. Nevertheless, removing the chattering effect is

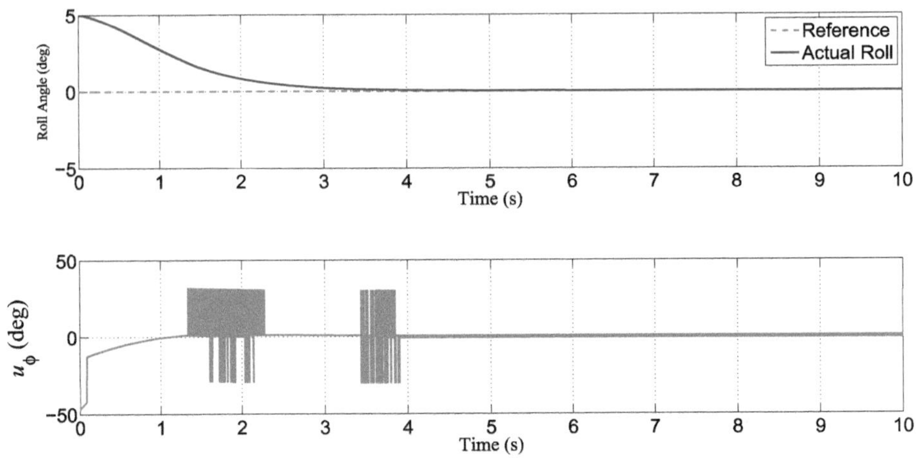

Figure 5.15 Action of the nested saturation controller with 2-SM on roll.

necessary to design a controller with high-order sliding mode (HOSM). The robust second-order differentiator is necessary for this controller too [48]; this robust differentiator is defined by

$$\dot{x}_0 = v_0 = -\lambda_0 |x_0 - s_l|^{2/3} \operatorname{sgn}(x_0 - s_l) + x_1, \tag{5.33}$$

$$\dot{x}_1 = v_1 = -\lambda_1 |x_1 - s_l|^{1/2} \operatorname{sgn}(x_1 - v_0) + x_2, \tag{5.34}$$

$$\dot{x}_2 = -\lambda_2 \operatorname{sgn} |x_2 - v_1|, \tag{5.35}$$

where x_0, x_1, and x_2 are real-time estimates of s_l, $\dot{s}_l$, and $\ddot{s}_l$, respectively, and $\lambda_0 = 15$, $\lambda_1 = 12$, and $\lambda_2 = 6.6$, are obtained by the theory which is shown in [48]. To obtain the nested saturation controller with high-order sliding mode we start with the controller designed in (5.29); thus, the nested saturation controller with high-order sliding mode obtained is

$$\begin{aligned} u_l = & - C_{1l} z_{2l} - \alpha_l [\ddot{s}_{1l} + 2(|\dot{s}_{1l}|^3 + |s_{1l}|^2)^{1/6} \operatorname{sgn}(\dot{s}_{1l} + |s_{1l}|^{2/3} \operatorname{sgn}(s_{1l}))] \\ & - \sigma_{2l} [k_{1l} z_{2l} + \sigma_{1l}(k_{1l} z_{1l})] / C_{2l}, \end{aligned} \tag{5.36}$$

where k_{1l} and α_l are nonnegative constants. The variables z_{1l} and z_{2l} have been defined in the design of the nested saturation controller. The values of the gains of the nested saturation controller with high-order sliding mode used in simulations are shown in Table 5.6. For the α_{x1l} gain fine-tuning is done in a heuristic form, which helps to reduce or remove the chattering effect.

In Fig. 5.16, the simulation results obtained for the nested saturation controller with high-order sliding mode are presented. In the upper graph of the figure, the target altitude of the fixed-wing UAV is achieved in less than 50 seconds, which is more time than with the other controllers. But as is appreciated in the lower graph of the same figure, the control signal presents a very low chattering effect after 30 seconds; it presents a weak chattering effect in comparison with the controllers based on sliding

Table 5.6 Gains of the nested saturation controller with HOSM.

Movement	a_{1l} [1/s^2]	k_{1l} [1/s]	k_{2l} [1/s]	α_{x1l} [s^2]
Altitude	4	0.32	0.104	0.08
Yaw	0.27	0.8	0.2	0.001
Roll	0.32	0.8	0.2	0.001

mode, nested saturation with sliding mode, and nested saturation with second-order sliding mode.

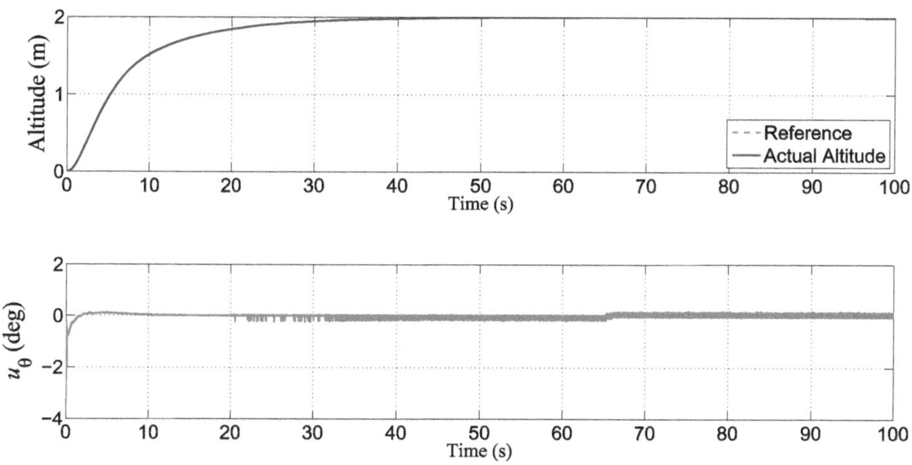

Figure 5.16 Action of the nested saturation controller with HOSM on altitude.

The simulation results for the yaw angle with the nested saturation controller with high-order sliding mode are shown in Fig. 5.17. It is observed that the yaw angle converges to the desired value in less than 10 seconds. In the lower graph of Fig. 5.17, it is observed that the chattering effect is reduced or almost eliminated when using the controllers by sliding mode, nested saturation with sliding mode, and nested saturation with second-order sliding mode.

As observed in the simulation results of Fig. 5.16 and Fig. 5.17, using the nested saturation controller with high-order sliding mode, we reduced or nearly eliminated the chattering effect for the roll angle. These results are shown in Fig. 5.18; in the upper graph it is observed that the control objective of bringing the roll angle to 0 degrees is achieved in less than 10 seconds; in the lower graph the elimination of the chattering effect in the control signal is observed.

5.7 Backstepping with SM

For the design of the backstepping controller with sliding mode [10], the system to control is given by

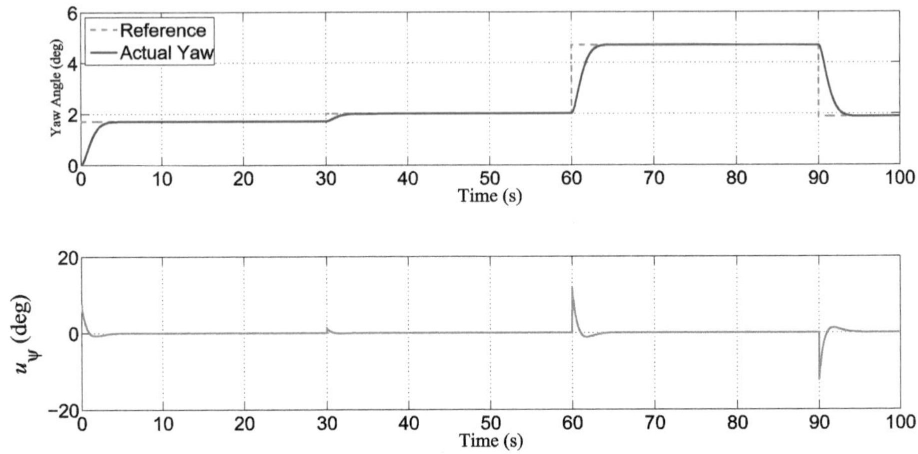

Figure 5.17 Action of the nested saturation controller with HOSM on yaw.

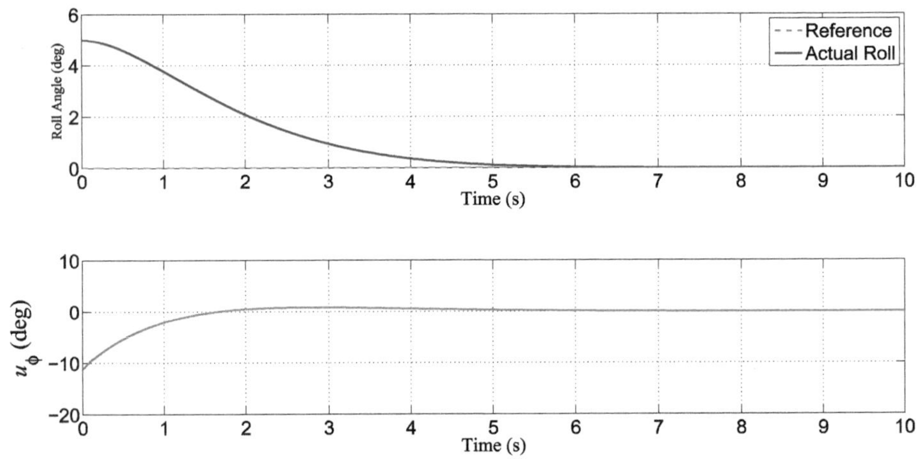

Figure 5.18 Action of the nested saturation controller with HOSM on roll.

$$\dot{x}_{1l} = x_{2l}, \tag{5.37}$$

$$\dot{x}_{2l} = C_{1l}x_{2l} + C_{2l}u_l. \tag{5.38}$$

As the first step to design the backstepping controller with sliding mode, it is necessary to define the error:

$$\tilde{e}_l = x_{1ld} - x_{1l}, \tag{5.39}$$

where x_{1ld} is the desired angular position and x_{1l} is the angular position of the fixed-wing UAV. As explained, $l \in \{\theta, \psi, \phi\}$ defines for which parameter the control law will be designed. It is defined as a Lyapunov candidate function with the error (5.39):

$$V_1 = \frac{1}{2}\tilde{e}_l^2. \tag{5.40}$$

Taking the derivative of (5.40) yields

$$\begin{aligned}\dot{V}_1 &= \tilde{e}_l\dot{\tilde{e}}_l &(5.41)\\ &= \tilde{e}_l(\dot{x}_{1ld} - \dot{x}_{1l}) &(5.42)\\ &= \tilde{e}_l(\dot{x}_{1ld} - x_{2l}), &(5.43)\end{aligned}$$

where x_{2l} is considered as a virtual control input, defined as

$$x_{2l} = \dot{x}_{1ld} + k_{1l}\tilde{e}_l - z_{1l}, \tag{5.44}$$

where z_{1l} is a new state variable and k_{1l} is a strictly nonnegative gain. Now, substituting x_{2l} in (5.43), we obtain

$$\dot{V}_1 = z_{1l}\tilde{e}_l - k_{1l}\tilde{e}_l^2. \tag{5.45}$$

Moreover, we design a sliding manifold $s = \dot{\tilde{e}}_l + k_l\tilde{e}_l = 0$, where k_l is a definite nonnegative gain. A Lyapunov candidate function is proposed for s, and besides the Lyapunov candidate function (5.40) is considered:

$$V_2 = V_1 + \frac{1}{2}s^2. \tag{5.46}$$

Now the derivative of (5.46) is obtained:

$$\begin{aligned}\dot{V}_2 &= \dot{V}_1 + s\dot{s} &(5.47)\\ &= k_{3l}e_1z_{1l} - k_{3l}k_{1l}e_l^2 + s(\dot{x}_{2ld} - \dot{x}_2 + k_{2l}(\dot{x}_{1ld} - \dot{x}_1)) &(5.48)\\ &= k_{3l}e_1z_{1l} - k_{3l}k_{1l}e_l^2 + s(\dot{x}_{2ld} - C_{1l} - C_{2l}u_l + k_{2l}(\dot{x}_{1ld} - x_{2l}) + k_{4l}e_lz_{1l}), &(5.49)\end{aligned}$$

because the system under analysis is described for only two states and the control input u_l appears in (5.18). Then, the backstepping control law with sliding mode is given by

$$u_l = \frac{\dot{x}_{2ld} + C_{1l}x_{2l} - k_{2l}(\dot{x}_{1ld} - x_{2l}) - k_{4l}e_lz_{1l}/s - k_{3l}s - \beta_{x1l}\,\mathrm{sgn}(s)}{C_{2l}}, \tag{5.50}$$

where k_{1l}, k_{2l}, k_{3l}, and β_{xl} are strictly nonnegative gains. Finally, the derivative of the Lyapunov candidate function (5.46) with the controller (5.50) is given by

$$\dot{V}_2 = -k_{3l}k_{1l}e_l^2 - k_{3l}s^2 - \beta_{x1l}\,\mathrm{sgn}(s) \le 0. \tag{5.51}$$

Table 5.7 shows the values of the gains obtained in a heuristic form of the backstepping controller with sliding mode.

Table 5.7 Gains of the backstepping controller with sliding mode.

Movement	k_{1l} [1/s]	k_{2l} [1/s]	k_{3l} [1/s]	k_{4l} [1/s^2]	β_{x1l} [1/s]
Altitude	1	5	0.9	0.1	1
Yaw	5	5	0.9	0.09	0.09
Roll	5	5	0.9	0.1	0.1

The obtained results for the altitude control with the backstepping controller with sliding mode are presented in Fig. 5.19. In the upper graph we observe that the altitude of the fixed-wing UAV achieves the reference in less than 8 seconds, and in the lower graph the control action applied with this controller is shown.

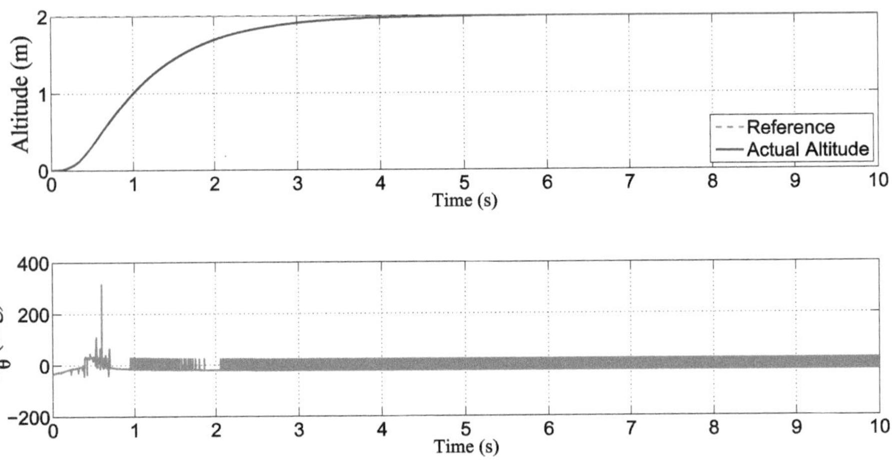

Figure 5.19 Action of the backstepping controller with sliding mode on altitude.

In the simulations of the yaw angle, a similar reference signal is considered as for the other controllers for this angle. The results obtained with the backstepping controller with sliding mode are presented in Fig. 5.20, where the upper graph shows that the yaw angle achieved the reference in less than 10 seconds. The control signal with the undesired chattering effect is observed in the lower graph, produced by sliding mode control.

In Fig. 5.21 the simulation results obtained for the roll angle using the backstepping controller with sliding mode are presented. In the upper graph of Fig. 5.21, it is observed that the objective to bring the roll angle to 0 degrees is achieved in less than 4 seconds. In the lower graph of the same Fig. 5.21 the control action is shown, which also presents the chattering effect.

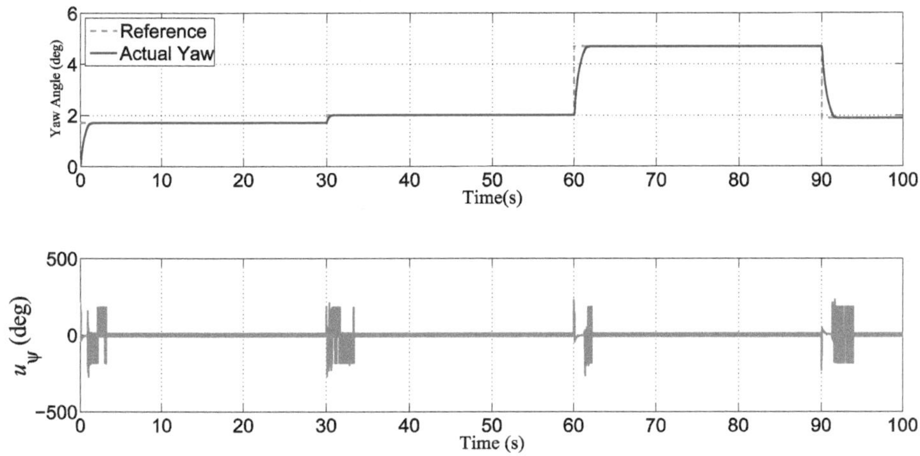

Figure 5.20 Action of the backstepping controller with sliding mode on yaw.

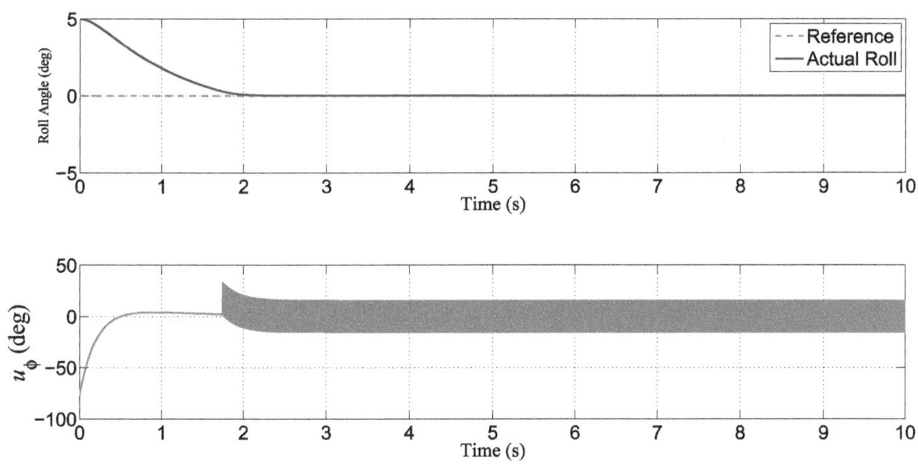

Figure 5.21 Action of the backstepping controller with sliding mode on roll.

5.8 Backstepping with 2-SM

For the design of the backstepping controller with second-order sliding mode, the same methodology for the obtention of the backstepping controller with sliding mode presented above was employed. With this controller the chattering effect is reduced, which is shown in the output of the sliding mode controllers. The difference is the addition of the derivative of the sliding manifold s with its respective gain and sign functions to the control law (5.50). This controller uses the same robust first-order differentiator given by Eqs. (5.30)–(5.31) and computes the derivative in real-time of the sliding manifold s. Then, considering the above, the backstepping controller with

Table 5.8 Gains of the backstepping controller with second-order sliding mode.

Movement	k_{1l} [1/s]	k_{2l} [1/s]	k_{3l} [1/s]	k_{4l} [1/s]	β_{x1l} [1/s]	β_{x2l} [s^2]
Altitude	0.01	10	5	0.1	0.5	0.5
Yaw	10	5	1	0.1	0.5	0.45
Roll	0.1	1	2.5	0.1	0.5	0.4552

second-order sliding mode obtained is

$$u_l = \frac{\dot{x}_{2ld} + C_{1l}x_{2l} - k_{2l}(\dot{x}_{1ld} - x_{2l}) - k_{4l}e_l z_{1l} - k_{3l}s - \beta_{x1l}\,\mathrm{sgn}(s) - \beta_{x1l}\,\mathrm{sgn}(\dot{s})}{C_{2l}}, \tag{5.52}$$

where k_{1l}, k_{2l}, k_{3l}, and β_{xl} are strictly nonnegative gains. The derivative of the Lyapunov candidate function (5.46), considering the obtained control law in (5.52), is

$$\dot{V}_2 = -k_{3l}k_{1l}e_l^2 - k_{3l}s^2 - \beta_{x1l}\,\mathrm{sgn}(s) - \beta_{x2l}\,\mathrm{sgn}(\dot{s}) \leq 0. \tag{5.53}$$

Table 5.8 shows the values of the gains in the backstepping controller with second-order sliding mode. The gains are tuned in a heuristic form, but aiming to reduce the chattering effect with the gains β_{x1l} and β_{x2l}.

The backstepping controller with second-order sliding mode for altitude presents a considerable reduction of the chattering effect compared with the backstepping controller with sliding mode (see the lower graph of Fig. 5.19 and Fig. 5.22). It can be observed that the desired reference was achieved within 8 seconds (see the upper graph of Fig. 5.22).

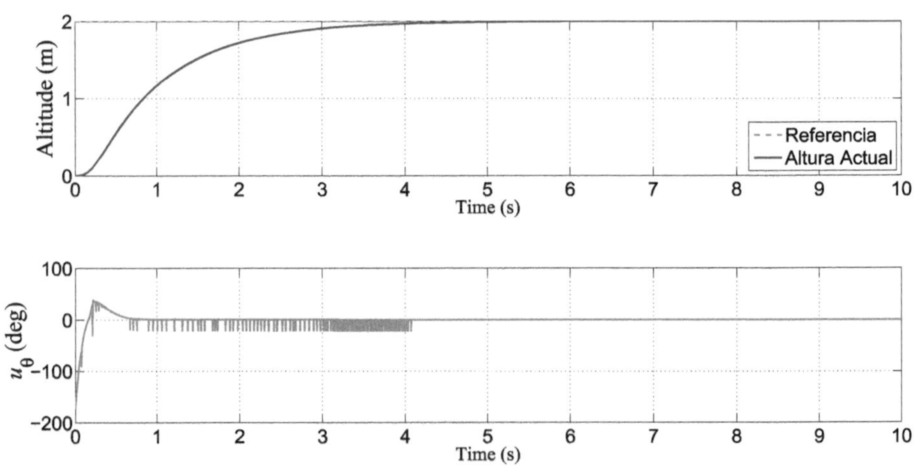

Figure 5.22 Action of the backstepping controller with second-order sliding mode on altitude.

In Fig. 5.23 the results obtained with the backstepping controller with second-order sliding mode for the yaw angle are shown. It is observed in the upper graph that the yaw angle converges to the desired reference in less than 10 seconds. In the lower graph of Fig. 5.23 the control action is shown, and it is observed that in comparison with the obtained results with the sliding mode controller and the backstepping controller with sliding mode, the backstepping controller with second-order sliding mode reduces the chattering effect.

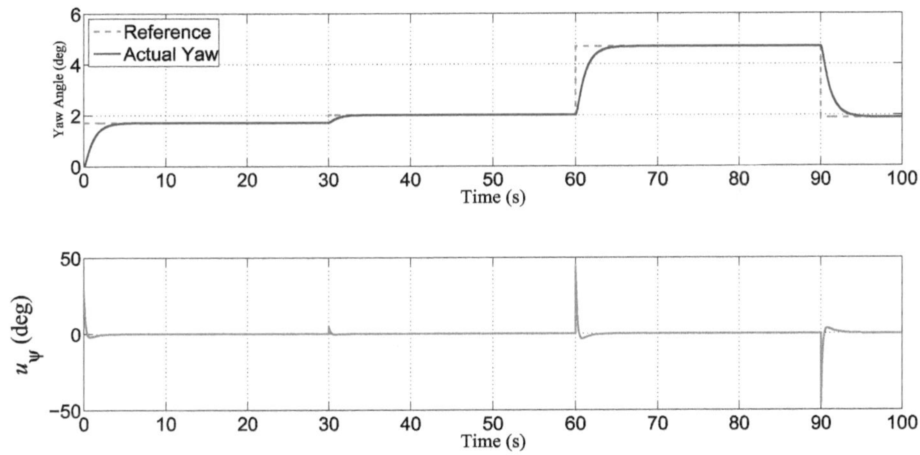

Figure 5.23 Action of the backstepping controller with second-order sliding mode on yaw.

The backstepping controller with second-order sliding mode applied for the roll angle shows a reduction of the chattering effect in the output of the controller (see the lower graph of Fig. 5.24). In the upper graph of the same Fig. 5.24, it is observed that the control objective is achieved in less than 7 seconds.

5.9 Backstepping with HOSM

The design of a backstepping controller with high-order sliding mode used the robust second-order differentiator given by (5.33)–(5.35). Second-order differentiation is used to obtain the real-time derivatives of the sliding manifold s. For this control the controller obtained with the backstepping methodology with sliding mode (5.50) is used. The backstepping controller with high-order sliding mode is given by

$$\begin{aligned} u_l = {} & [-C_{1l}x_{2l} - k_{2l}(\dot{x}_{1ld} - x_{2l}) - k_{4l}z_{2l}e_l - k_{3l}s \\ & - \alpha_l[\ddot{s}_{1l} + 2(|\dot{s}_{1l}|^3 + |s_{1l}|^2)^{1/6}\operatorname{sgn}(\dot{s}_{1l} + |s_{1l}|^{2/3}\operatorname{sgn}(s_{1l}))]]/C_{2l}, \end{aligned} \tag{5.54}$$

where k_{1l}, k_{2l}, k_{3l}, k_{4l}, and α_l are strictly nonnegative gains. The gains of the backstepping controller with high-order sliding mode are shown in Table 5.9. For the gains

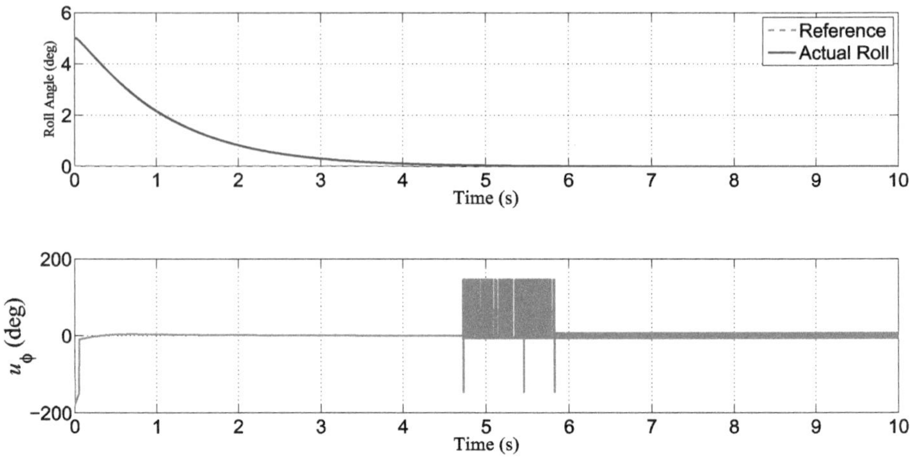

Figure 5.24 Action of the backstepping controller with second-order sliding mode on roll.

Table 5.9 Gains of the backstepping controller with HOSM.

Movement	k_{1l} [1/s]	k_{2l} [1/s]	k_{3l} [1/s]	k_{4l} [1/s]	α_{x1l} [s^2]
Altitude	0.001	11	5	0.1	0.0001
Yaw	0.01	1	5	0.01	0.0001
Roll	0.001	1	5	0.1	0.0001

of this controller, the gain α_{x1l} is fine-tuned. The other gains implicit in the controller are tuned in a heuristic form.

The derivative of the Lyapunov candidate function (5.46) with the controller (5.54) is given by

$$\dot{V}_2 = -k_{3l}k_{1l}e_l^2 - k_{3l}s^2 - \alpha_l[\ddot{s}_{1l} + 2(|\dot{s}_{1l}|^3 + |s_{1l}|^2)^{1/6}\,\mathrm{sgn}(\dot{s}_{1l} + |s_{1l}|^{2/3}\,\mathrm{sgn}(s_{1l}))] \leq 0. \tag{5.55}$$

Fig. 5.25 shows the simulation results for the backstepping controller with high-order sliding mode for altitude. In the lower graph of Fig. 5.25 a considerable reduction of the chattering effect is appreciated, and in the upper graph of Fig. 5.25 it is observed that the altitude converges to the desired reference in less than 6 seconds.

We used the same reference signal for the yaw angle as for the other controllers. In the upper graph of Fig. 5.26 it is observed that the control objective is achieved in less than 10 seconds, and in the lower graph the control action with the reduced chattering effect is presented.

In the simulation for the roll angle, the fixed-wing UAV has an initial angle of 5 degrees. The objective of the backstepping controller with high-order sliding mode is to bring the roll angle to 0 degrees, while reducing or eliminating the chattering effect of the controller output. The upper graph of Fig. 5.27 shows that the roll angle achieves

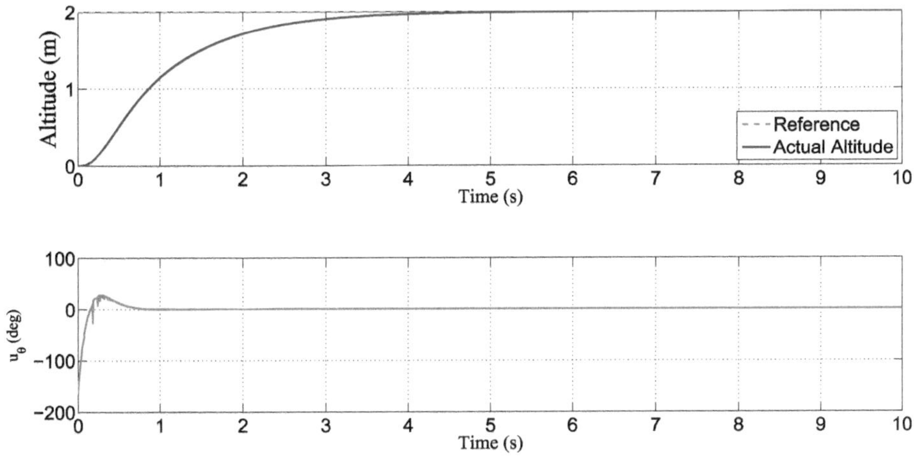

Figure 5.25 Action of the backstepping controller with HOSM on altitude.

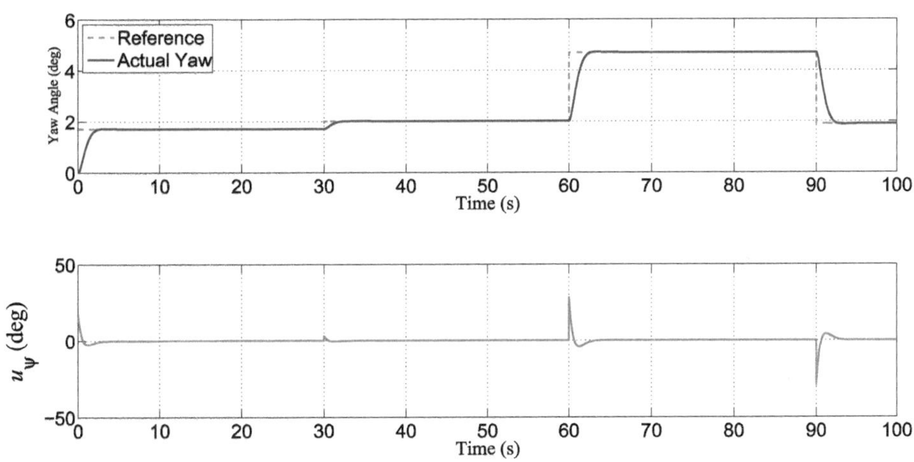

Figure 5.26 Action of the backstepping controller with HOSM on yaw.

the reference in less than 8 seconds, and in the lower graph of Fig. 5.27 the control action with the chattering effect almost zero is shown.

In Table 5.10 the results obtained with the $\mathcal{L}_2$ norm are shown. It is observed that the nonlinear controller which presents a low error in altitude is the backstepping controller, and the nested saturation methodology presents a lower effort in the controller signal, but the nested saturation controller presents a more significant error than the backstepping controller and the sliding mode controller. For the yaw angle control, the nonlinear controller that presents the lowest error is the sliding mode controller. Still, this controller presents a more significant effort in the control signal and has a chattering effect in the controller signal. The nested saturation controller for the yaw angle presents a lower effort, but it has a bigger error than the backstepping controller

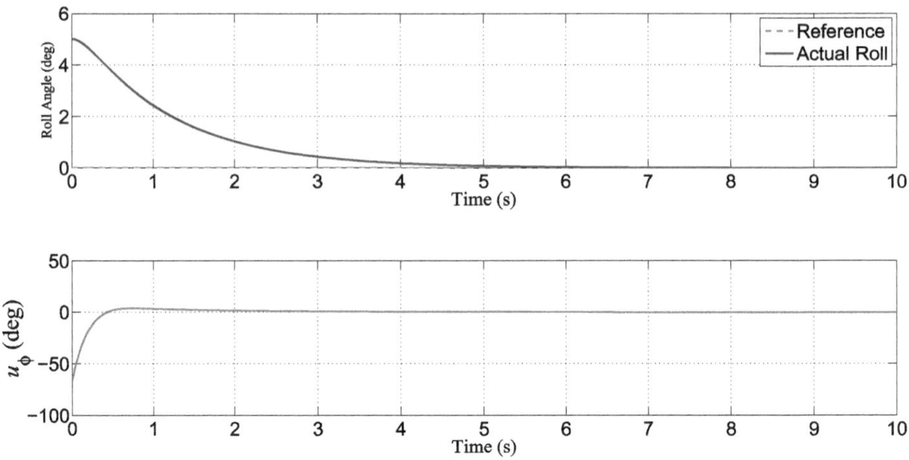

Figure 5.27 Action of the backstepping controller with HOSM on roll.

and the sliding mode controller. The sliding mode controller presented a lower error for the roll angle, besides a lower controller effort.

In the combination of nonlinear controllers, the controller that presented the lowest error in altitude was the backstepping controller with sliding mode. Still, this combination generates a more significant controller effort than the other combinations of nonlinear controllers. The backstepping controller with second-order sliding mode presented a lower error in the yaw angle, besides presenting a lower controller effort than the other combinations of nonlinear controllers. The backstepping controller with sliding mode presented a lower error in the roll angle. However, this combination of nonlinear controllers presented a more significant effort for roll angle control.

5.10 MIT rule based on the gradient method with sliding mode theory

We have designed a PD control law with adaptive gains; thus, the adaptive part of the controller is given by the proportional and the derivative gains. These gains are defined as $\hat{k}_{pla}$ and $\hat{k}_{vla}$, respectively.

The methodology to design the adaptive control is based on the MRAS, to design the adjustment mechanism by the MIT rule based on the gradient method with sliding mode theory. We have modified the MIT rule based on the gradient method by inserting the theory of first-order sliding mode, second-order sliding mode (2-SM), and high-order sliding mode (HOSM). Then, we first obtain a robust adjustment mechanism, stabilizing the system or the fixed-wing UAV in the presence of unknown perturbations like wind gusts and trying to reduce the chattering effect.

The block diagram representing the proposed MRAS is shown in Fig. 5.28, where the *Plant* is the aerodynamic model which describes the fixed-wing MAV and the block called *Model* represents the model reference.

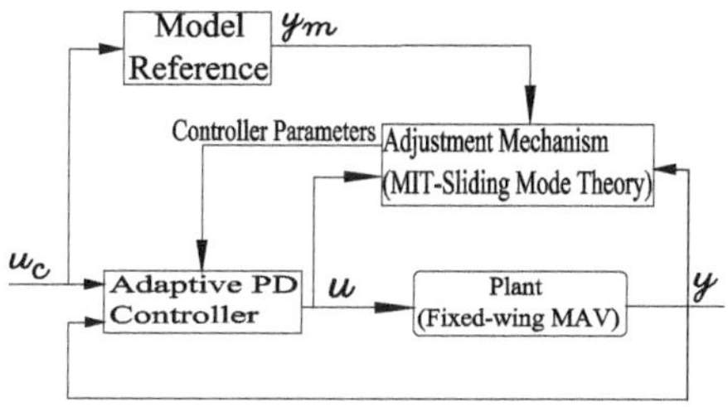

Figure 5.28 Block diagram of the MRAS applied to a fixed-wing UAV.

Consider Eqs. (5.27)–(5.28). The adaptive control is given by

$$u_l = \hat{k}_{pla} e_l + \hat{k}_{vla} \dot{e}_l, \tag{5.56}$$

where $l \in \{\theta, \psi, \phi\}$, that is, $u_\theta = \delta_e$, $u_\psi = \delta_r$, and $u_\phi = \delta_a$, and $\hat{k}_{pla}$ and $\hat{k}_{vla}$ are called the position and velocity gains, respectively; thus, these are the adaptive gains. The error of the directional and lateral dynamics has been defined as $e_l = x_{1l} - x_{1l}^d$. The gains of the PD control have a subscript to indicate the adjustment mechanism which applies, that is, $a = a_1, a_2, a_3, a_4$, where:

- a_1 corresponds to the MIT rule,
- a_2 corresponds to the MIT rule with sliding mode (MIT-SM),
- a_3 uses the MIT rule with second-order sliding mode (MIT-2SM), and
- a_4 represents the MIT rule with high-order sliding mode (MIT-HOSM).

Therefore, for the design of the MIT rule with sliding mode theory, let us define a cost function given by

$$J = \frac{1}{2} e_{l_m}^2(t), \tag{5.57}$$

where

$$e_{l_m} = x_{1l_m} - x_{1l}, \tag{5.58}$$

where x_{1l_m} is the output from the reference model. We have followed the methodology that has been presented in [3] for the MIT rule. Then, considering [3], the aerodynamic model has been transformed into the representation of a transference function to develop the derivatives of sensitivity; these have been obtained by computing partial derivatives with respect to the controller parameters $\hat{k}_{pla}$ and $\hat{k}_{vla}$. Thus, the closed-loop transfer function of the inner loop with the adaptive PD controller has been defined as

$$x_{1l} = \frac{C_{2l}(\hat{k}_{pl} + \hat{k}_{vl} s)}{s^2 + (C_{1l} + C_{2l}\hat{k}_{vl})s + C_{2l}\hat{k}_{pl}} x_{1l}^d, \tag{5.59}$$

and the model of reference for the fixed-wing UAV dynamics has been defined as

$$x_{1lm} = \frac{\omega_n^2}{s^2 + 2\zeta\omega_n s + \omega_n^2} x_{1l}^d, \tag{5.60}$$

where $\zeta = 3.17$ and $\omega = 3.16$. We consider (5.58)–(5.60) and calculate the partial derivatives with respect to $\hat{k}_{pla}$ and $\hat{k}_{vla}$. We obtain

$$\frac{\partial e_{l_m}}{\partial \hat{k}_{pl}} = \frac{C_{2l}}{s^2 + (C_{1l} + C_{2l}\hat{k}_{vl})s + C_{2l}\hat{k}_{pl}} (x_{1l} - x_{1l}^d), \tag{5.61}$$

$$\frac{\partial e_{l_m}}{\partial \hat{k}_{vl}} = \frac{C_{2l}s}{s^2 + (C_{1l} + C_{2l}\hat{k}_{vl})s + C_{2l}\hat{k}_{pl}} (x_{1l}). \tag{5.62}$$

Generally, expressions (5.61) and (5.62) cannot be used due to the unknown parameters $\hat{k}_{pla}$ and $\hat{k}_{vla}$ [3]. Thus, an optimum case has been assumed, defined as

$$s^2 + (C_{1l} + C_{2l}\hat{k}_{vl})s + C_{2l}\hat{k}_{pl} = s^2 + 2\zeta\omega_n s + \omega_n^2. \tag{5.63}$$

Thus, after these approximations, we obtain the differential equations to adjust the gains of the adaptive PD controller:

$$\dot{\hat{k}}_{pla_1} = -\gamma_{1l} \left(\frac{1}{s^2 + 2\zeta\omega_n s + \omega_n^2} (x_{1l} - x_{1l}^d) \right) e_{l_m}, \tag{5.64}$$

$$\dot{\hat{k}}_{vla_1} = -\gamma_{2l} \left(\frac{s}{s^2 + 2\zeta\omega_n s + \omega_n^2} (x_{1l}) \right) e_{l_m}. \tag{5.65}$$

Now, an MIT rule with first-order sliding mode is proposed; this approach is different from the one defined in [3]. Then, a sliding mode surface is defined as $\sigma_l = \dot{x}_{1lm} - x_{2l} + k_{1l}e_{l_m}$ (with this formulation we search to increase the stability and robustness in the adjustment mechanism), where $k_{1l} > 0$. Then, the differential equations of the adaptive controller with sliding mode methodology are given by

$$\dot{\hat{k}}_{pla_2} = -\gamma_{1l} \left(\frac{1}{s^2 + 2\zeta\omega_n s + \omega_n^2} (x_{1l} - x_{1l}^d) \right) (\beta_{p1l} \operatorname{sgn}(\sigma_l)), \tag{5.66}$$

$$\dot{\hat{k}}_{vla_2} = -\gamma_{2l} \left(\frac{s}{s^2 + 2\zeta\omega_n s + \omega_n^2} (x_{1l}) \right) (\beta_{v1l} \operatorname{sgn}(\sigma_l)), \tag{5.67}$$

where the gains are $\beta_{p1l}, \beta_{v1l} > 0$. Due to the chattering effect of the first-order sliding mode, let us design an adjustment mechanism with a second-order sliding mode. Even this formulation is different from the one presented in [3]. This second-order sliding mode includes a robust differentiator of first order (5.33). Thus, the differential equations of the adaptive PD controller with second-order sliding mode are defined as

$$\dot{\hat{k}}_{pla_3} = -\gamma_{1l} \left(\frac{1}{s^2 + 2\zeta\omega_n s + \omega_n^2} (x_{1l} - x_{1l}^d) \right) (\beta_{p1l} \operatorname{sgn}(\sigma_l) + \beta_{p2l} \operatorname{sgn}(\dot{\sigma}_l)), \tag{5.68}$$

$$\dot{\hat{k}}_{vla_3} = -\gamma_{2l}\left(\frac{s}{s^2+2\zeta\omega_n s+\omega_n^2}(x_{1l})\right)(\beta_{v1l}\,\mathrm{sgn}(\sigma_l)+\beta_{v2l}\,\mathrm{sgn}(\dot{\sigma}_l)), \tag{5.69}$$

where the gains are $\beta_{p1l}, \beta_{p2l}, \beta_{v1l}, \beta_{v2l} > 0$. To reduce or eliminate the chattering effect in the second-order sliding mode, we have designed an adjustment mechanism with HOSM. To design the adjustment mechanism, a robust differentiator of second order (5.33). Finally, the differential equations of the adaptive PD controller based on the MIT rule by the gradient method with high-order sliding mode (HOSM) are defined by

$$\begin{aligned}\dot{\hat{k}}_{pla_4} &= -\gamma_{1l}\left(\frac{1}{s^2+2\zeta\omega_n s+\omega_n^2}(x_{1l}-x_{1l}^d)\right)\\ &\quad(\alpha_{pl}[\ddot{\sigma}_l+2(|\dot{\sigma}_l|^3+|\sigma_l|^2)^{1/6}\,\mathrm{sgn}(\dot{\sigma}_l+|\sigma_l|^{2/3}\,\mathrm{sgn}(\sigma_l))]), \qquad (5.70)\\ \dot{\hat{k}}_{vla_4} &= -\gamma_{2l}\left(\frac{s}{s^2+2\zeta\omega_n s+\omega_n^2}(x_{1l})\right)\\ &\quad(\alpha_{vl}[\ddot{\sigma}_l+2(|\dot{\sigma}_l|^3+|\sigma_l|^2)^{1/6}\,\mathrm{sgn}(\dot{\sigma}_l+|\sigma_l|^{2/3}\,\mathrm{sgn}(\sigma_l))]), \qquad (5.71)\end{aligned}$$

where the gains are $\alpha_{pl}, \alpha_{vl} > 0$.

Fig. 5.29 shows the results obtained with the different adaptive mechanisms designed for the adaptive PD controller. The error with the mechanism based on MIT-HOSM is smaller than those with the mechanisms based on MIT, MIT-SM, and MIT-2SM. It is appreciated that the adaptive controller achieves the desired altitude of 10 meters in a critically damped form; see Fig. 5.29.

The control signals generated by the adaptive PD with the different adjustment mechanisms are presented in Fig. 5.30. The control effort of the adaptive mechanism based on MIT-HOSM is more significant than the one generated by the MIT-2SM mechanism. But, the adaptive mechanism based on MIT-HOSM is smaller than the MIT and MIT-2SM mechanisms (see Table 5.10).

The mechanism that presented a more significant error in altitude was MIT-2SM. The control effort by the mechanism based on MIT-2SM is smaller than that by MIT, MIT-SM, and MIT-HOSM (see Table 5.10).

In Fig. 5.31 the minimization of the cost function with the different adaptive mechanisms is presented. It is proved in the figure that $J(\theta) \to 0$. In Fig. 5.32 the performance of the adaptive proportional gain $\hat{k}_p$ is presented, and in Fig. 5.33 the performance of the adaptive derivative gain $\hat{k}_v$ for altitude with the different adaptive mechanisms is appreciated. We have done a zoom or a reduction in the numerical scale on the control signal, and it is appreciated how the perturbations affect the control signal presenting a significant reduction of such unknown perturbations and the reduction of the chattering effect shown in the sliding mode (see Fig. 5.34). Thus, with the adaptive mechanism based on MIT with high-order sliding mode, a better flight at high altitudes is obtained in the presence of unknown perturbations as wind gusts.

The response of the sliding manifold for the adjustment mechanism based on sliding mode theory for the adaptive PD controller for altitude is presented in Fig. 5.35 for the adaptive proportional gain and in Fig. 5.36 for the adaptive derivative gain.

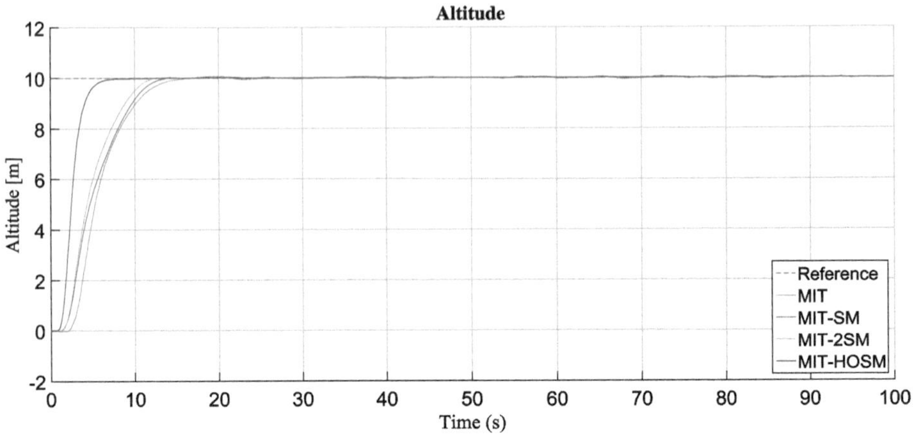

Figure 5.29 Adaptive PD controller response for altitude (with disturbances).

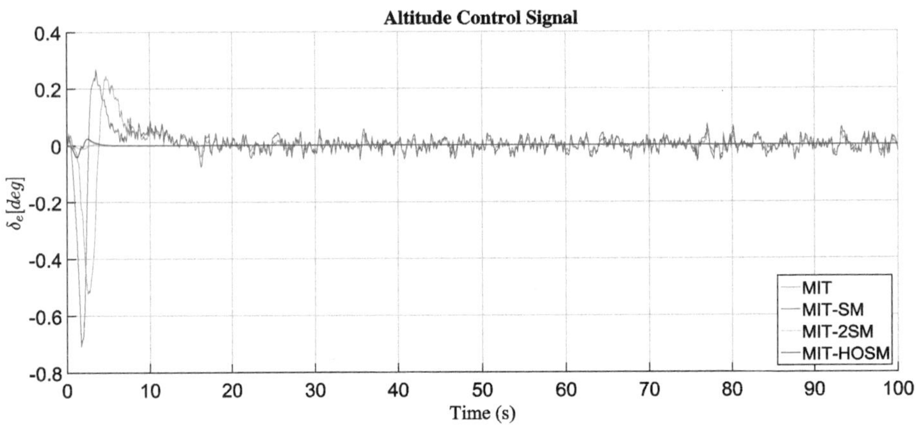

Figure 5.30 Control signal of the adaptive PD controller for altitude (with disturbances).

Fig. 5.37 shows the results obtained with the PD adaptive controller with the different adjustment mechanisms. The MIT-HOSM mechanism is smaller than the adjustment mechanisms based on MIT, MIT-SM, and MIT-HOSM. On the other hand, the control effort of the adaptive mechanism MIT-HOSM is more significant than the adjustment mechanism based on MIT, MIT-SM, and MIT-HOSM (see Table 5.10). In Fig. 5.38 the control response of the adaptive PD controller with the different adjustment mechanisms is presented.

The adaptive mechanism based on MIT-2SM presented a more significant error in the yaw angle than MIT, MIT-SM, and MIT-HOSM. On the other hand, the control effort by the PD adaptive control with the adjustment mechanism based on MIT-2SM is more significant than that with the adjustment mechanism based on MIT-SM or MIT. But, the control signal with the adjustment mechanism based on MIT-2SM is more

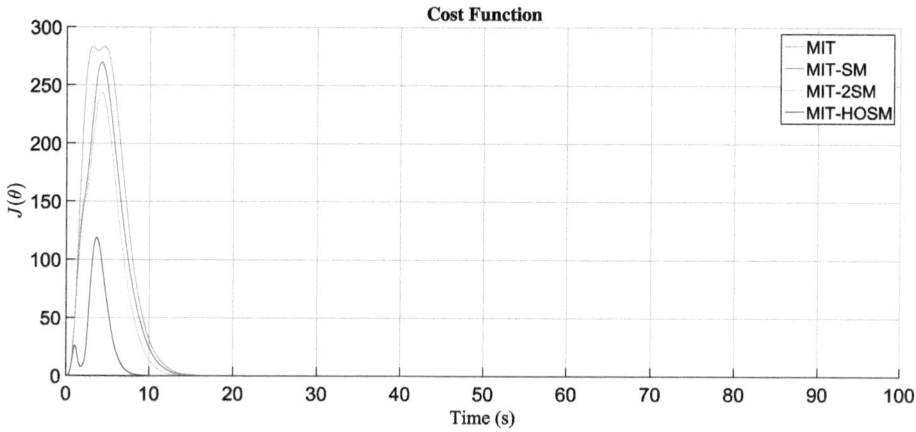

Figure 5.31 Minimization of the cost function for altitude (with disturbances).

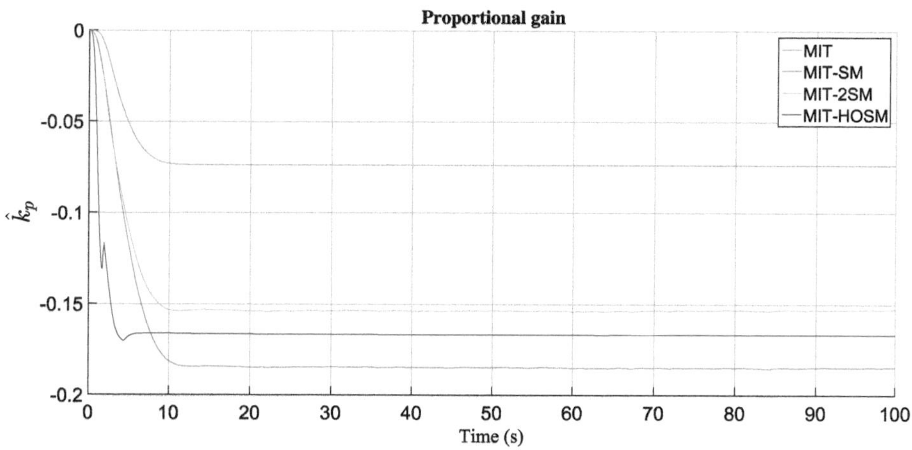

Figure 5.32 Response of the adaptive proportional gain controller for altitude (with disturbances).

significant than that based on MIT-HOSM; see Table 5.10. In Fig. 5.39 the result of the minimization of the cost function for the yaw angle is presented. It is appreciated that $J(\psi) \to 0$. In Fig. 5.40 the performance of the adaptive proportional gain $\hat{k}_p$ is shown, and in Fig. 5.41 the performance of the adaptive derivative gain $\hat{k}_v$ for yaw angle for the different adaptive mechanisms is appreciated.

In Fig. 5.42 a zoom of the control signals generated by the adaptive PD controller with different adjustment mechanisms is presented. Some reduction of the unknown perturbations and a small attenuation of the chattering effect are appreciated.

The response of the sliding manifold for the adjustment mechanism based on sliding mode theory for the adaptive PD controller for the yaw angle is presented in

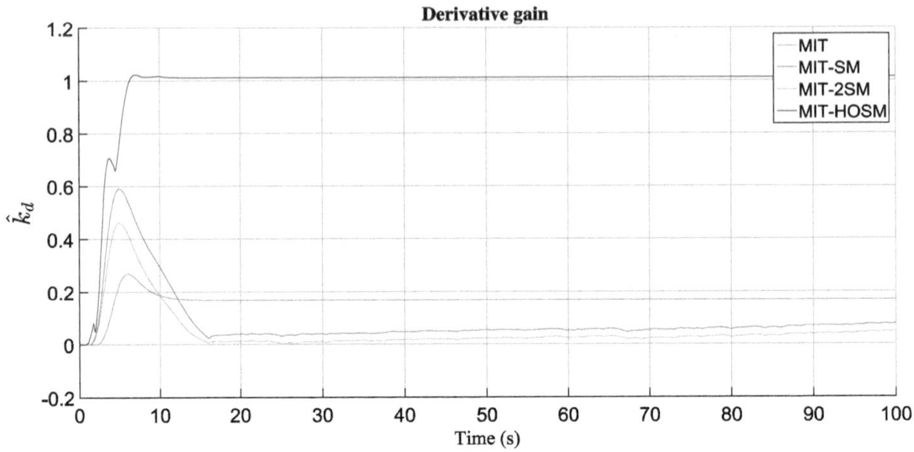

Figure 5.33 Response of the adaptive derivative gain controller for altitude (with disturbances).

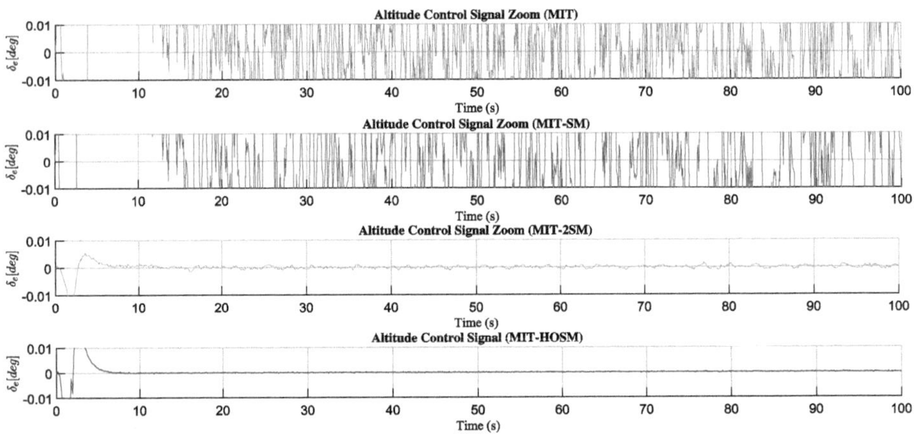

Figure 5.34 Zoom of the control signal of the adaptive PD controller for altitude (with disturbances).

Fig. 5.43 for the adaptive proportional gain and in Fig. 5.44 for the adaptive derivative gain.

In Fig. 5.45 the response of the adaptive PD controller with the different adjustment mechanism is presented. It is appreciated that the MIT-HOSM adjustment mechanism is smaller than the adjustment mechanism based on MIT, MIT-SM, and MIT-2SM, respectively. Also, the control effort of the adaptive PD controller with the adjustment mechanism based on MIT-HOSM is more significant than the adjustment mechanism based on MIT, MIT-SM, and MIT-2SM, respectively; see Table 5.10.

The adjustment mechanism based on the MIT rule is presented in [3], the error in roll angle is more significant than that with the adjustment mechanism based on MIT-

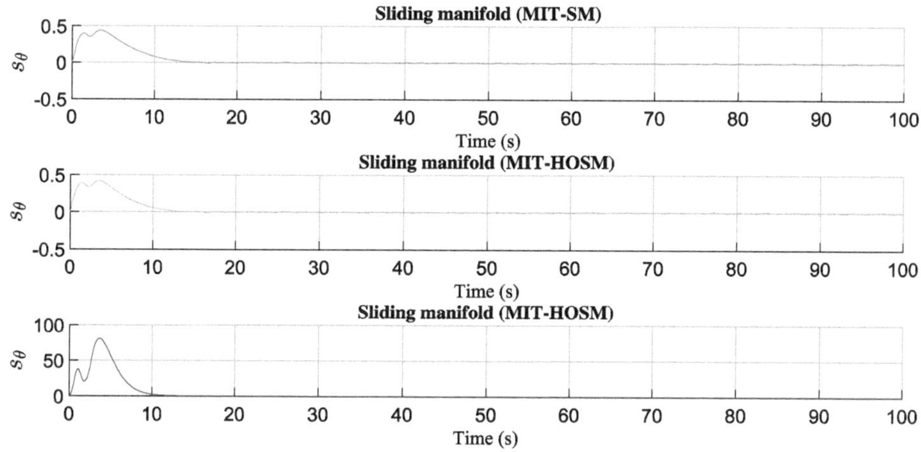

Figure 5.35 Response of the sliding manifold controller with adaptive proportional gain for altitude (with disturbances).

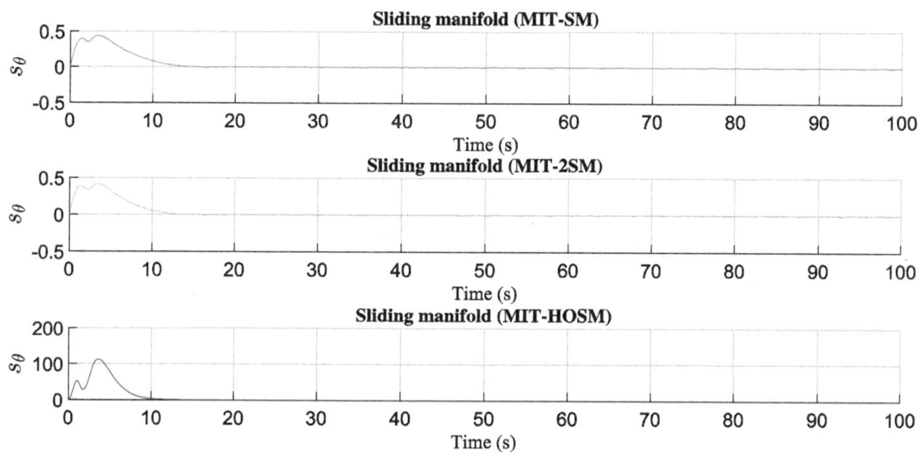

Figure 5.36 Response of the sliding manifold controller with adaptive derivative gain for altitude (with disturbances).

SM, MIT-2SM, and MIT-HOSM. The control effort of the adaptive PD controller with the adjustment mechanism based on the MIT as proposed in [3] is more significant than that of the adjustment mechanism based on MIT-2SM and MIT-HOSM. Still, it is smaller than that of the adjustment mechanism based on MIT-SM. Fig. 5.46 presents the adaptive PD controller with a different adaptive mechanism.

The minimization of the cost function for the roll angle is presented in Fig. 5.47. It is appreciated that $J(\phi) \rightarrow 0$. Also, in Fig. 5.48 the performance of the adaptive proportional gain $\hat{k}_p$ is shown, and in Fig. 5.49 the performance of the adaptive derivative gain $\hat{k}_v$ for the roll angle for the different adjustment mechanisms is appreciated.

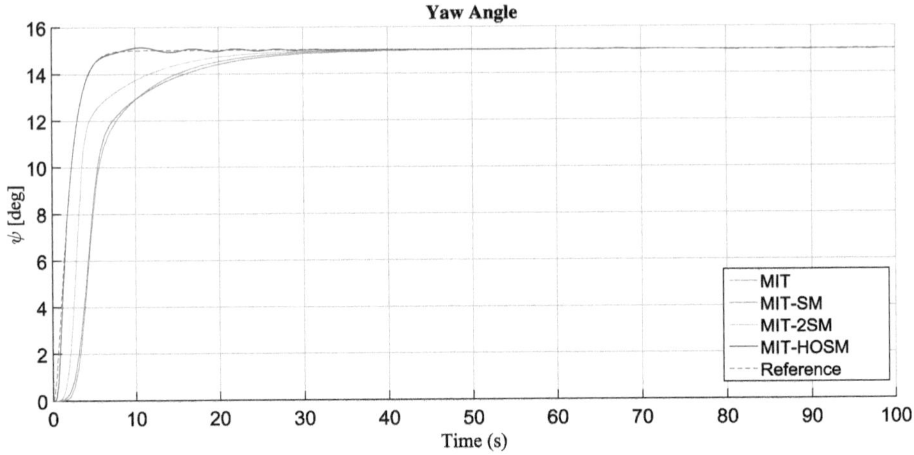

Figure 5.37 Adaptive PD controller response for the yaw angle (with disturbances).

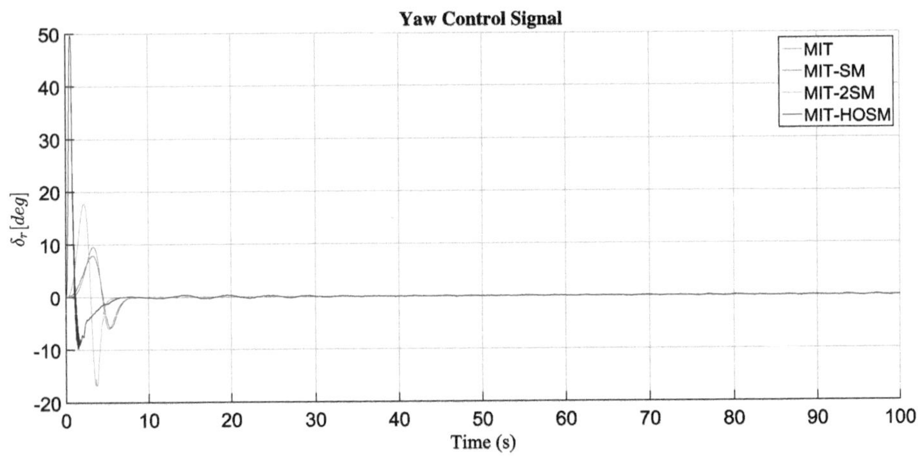

Figure 5.38 Control signal of the adaptive PD controller for the yaw angle (with disturbances).

Fig. 5.50 presents a zoom of the control signals generated by the adaptive PD controller with the different adjustment mechanisms, and a significant reduction of the perturbations is appreciated. Even a considerable attenuation of the chattering effect is achieved.

The response of the sliding manifold for the adjustment mechanism based on sliding mode theory for the adaptive PD controller for the yaw angle is presented in Fig. 5.51 for the adaptive proportional gain and in Fig. 5.52 for the adaptive derivative gain.

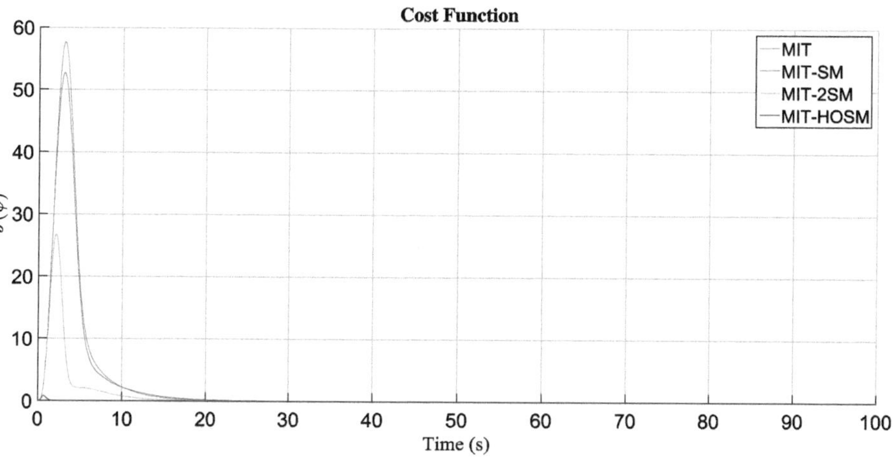

Figure 5.39 Minimization of the cost function for the yaw angle (with disturbances).

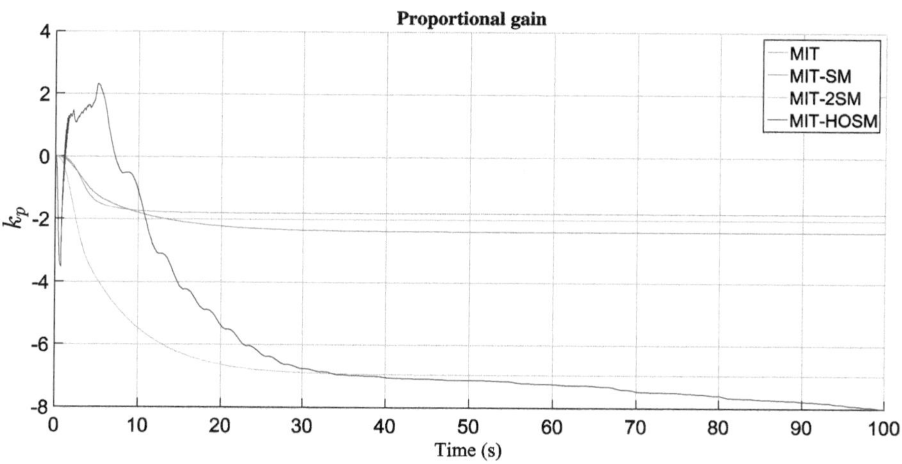

Figure 5.40 Response of the adaptive proportional gain for the yaw angle (with disturbances).

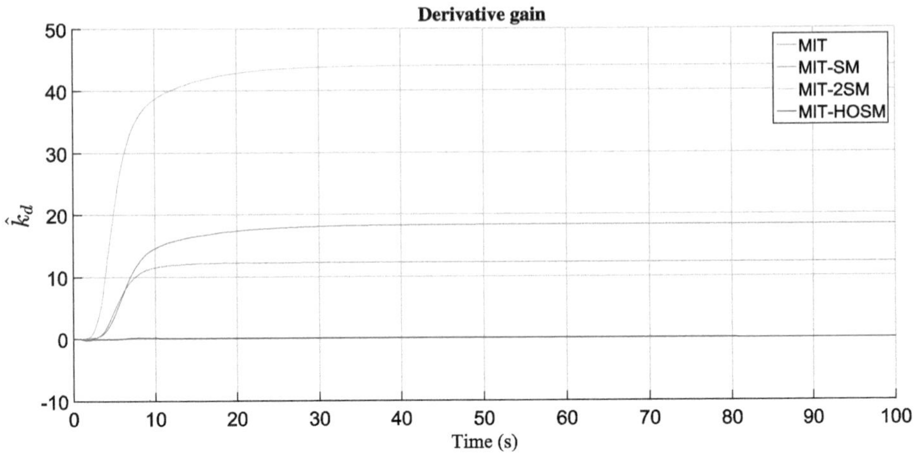

Figure 5.41 Response of the adaptive derivative gain for the yaw angle (with disturbances).

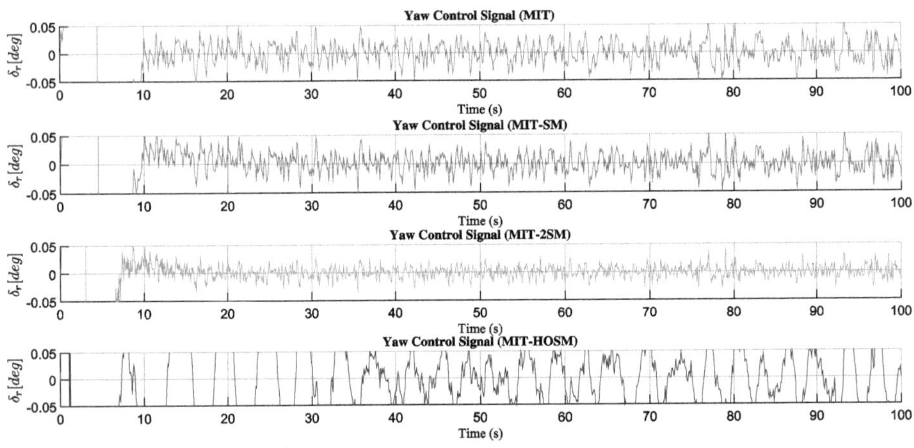

Figure 5.42 Zoom of the control signal of the adaptive PD controller for the yaw angle (with disturbances).

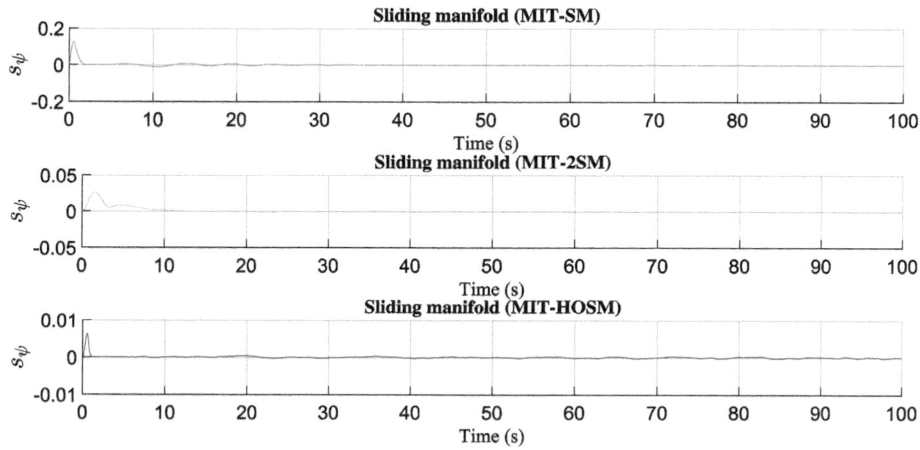

Figure 5.43 Response of the sliding manifold in the adaptive proportional gain for the yaw angle (with disturbances).

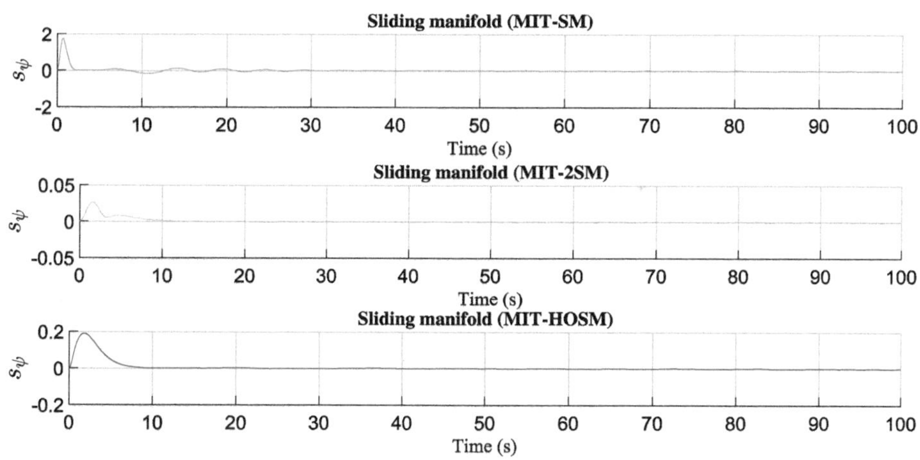

Figure 5.44 Response of the sliding manifold in the adaptive derivative gain for the yaw angle (with disturbances).

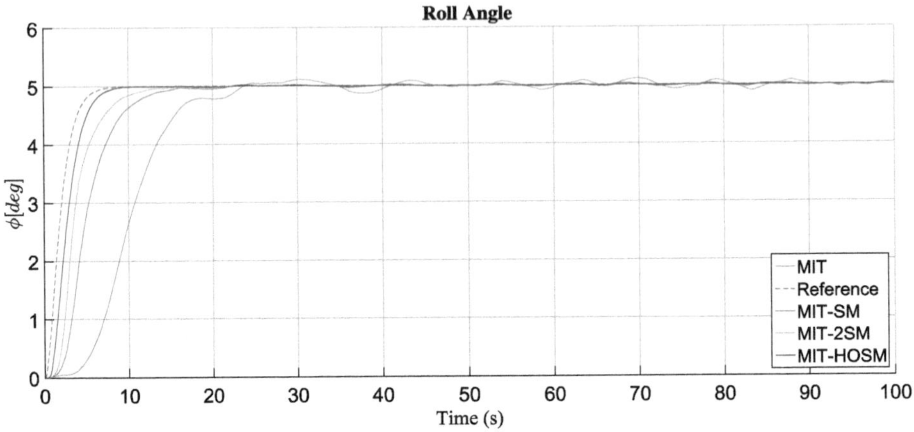

Figure 5.45 Adaptive PD controller response for the roll angle (with disturbances).

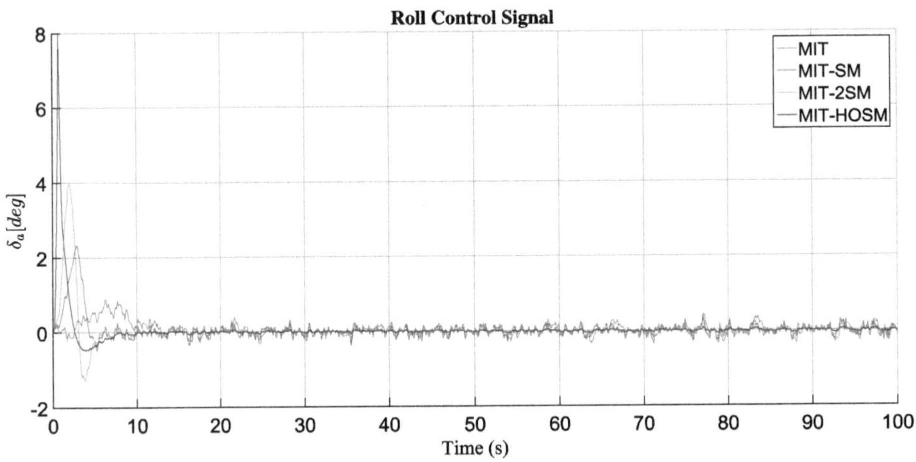

Figure 5.46 Control signal of the adaptive PD controller for the roll angle (with disturbances).

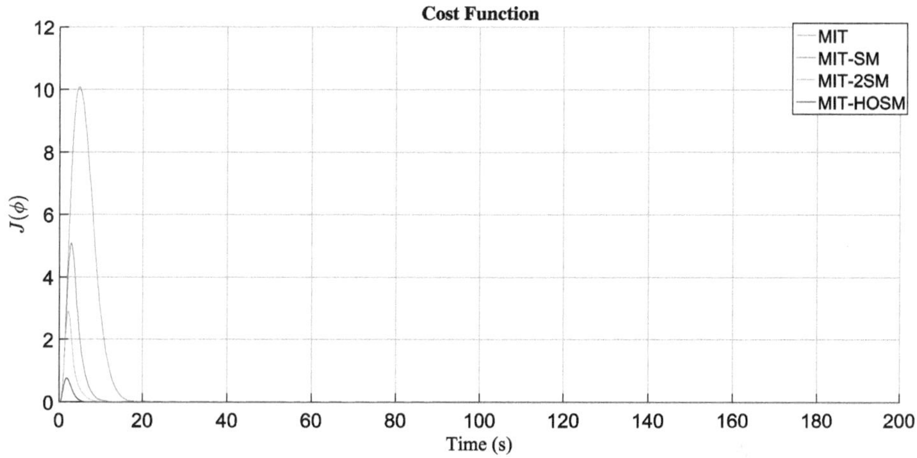

Figure 5.47 Minimization of the cost function for the roll angle (with disturbances).

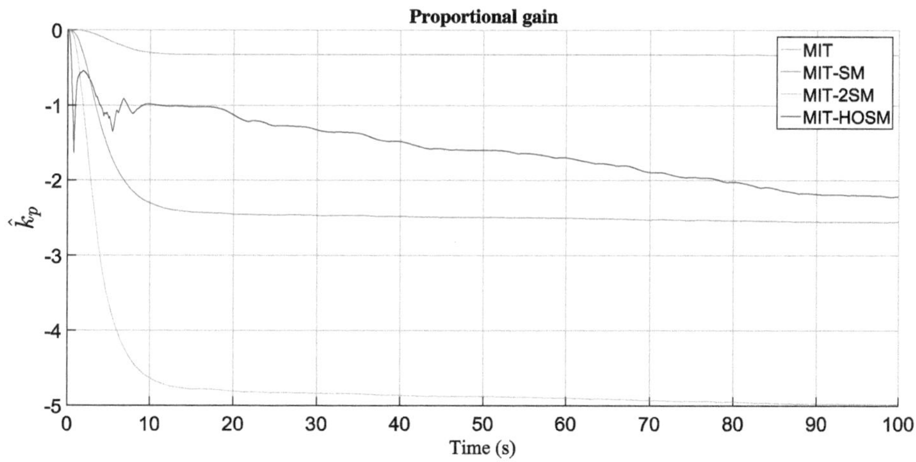

Figure 5.48 Response of the adaptive proportional gain for the roll angle (with disturbances).

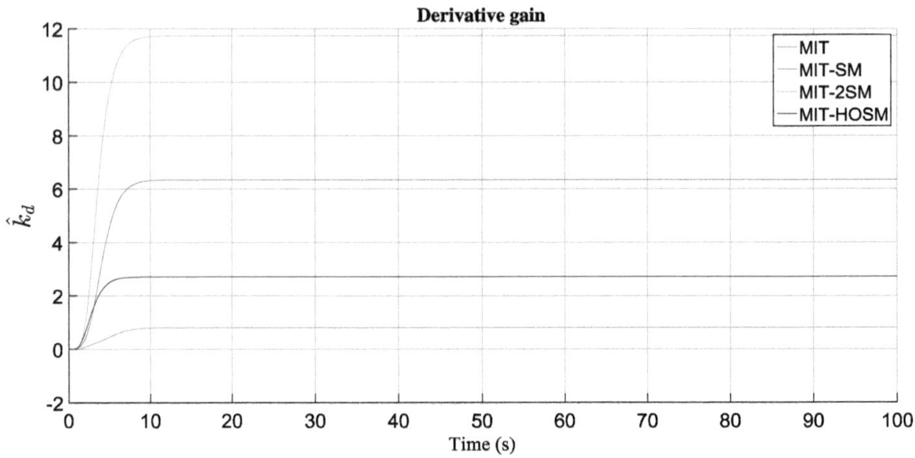

Figure 5.49 Response of the adaptive derivative gain for the roll angle (with disturbances).

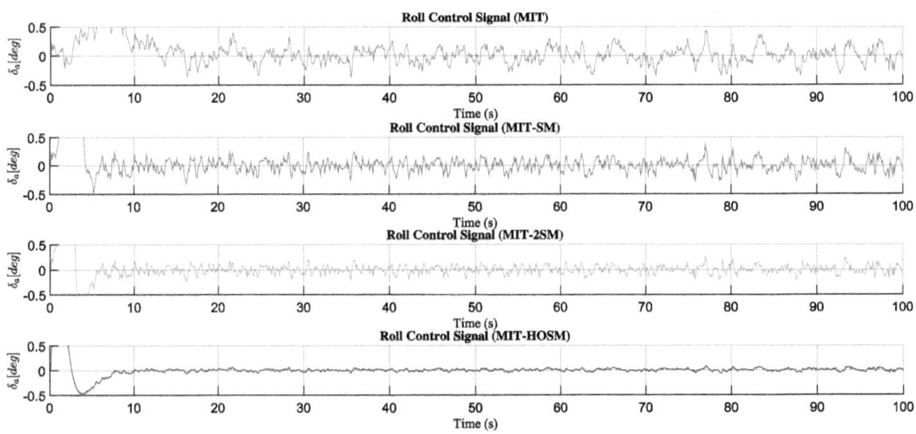

Figure 5.50 Zoom of the control signal of the adaptive PD controller for the roll angle (with disturbances).

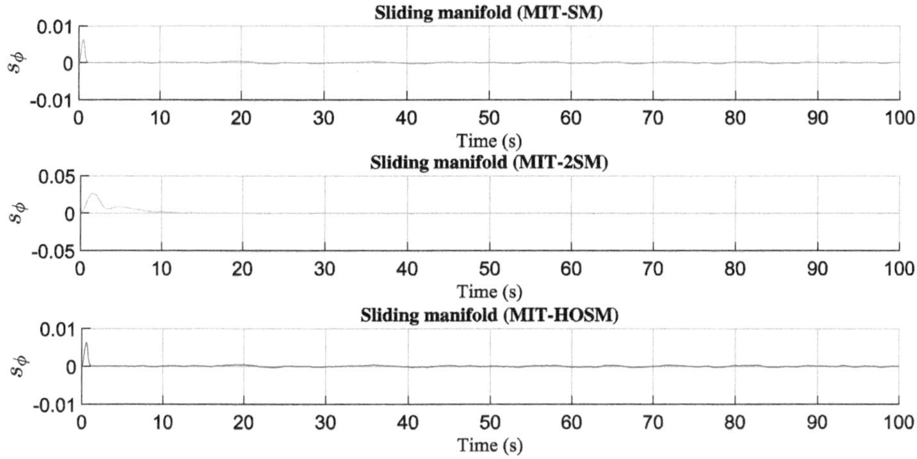

Figure 5.51 Response of the sliding manifold in the adaptive proportional gain for the roll angle (with disturbances).

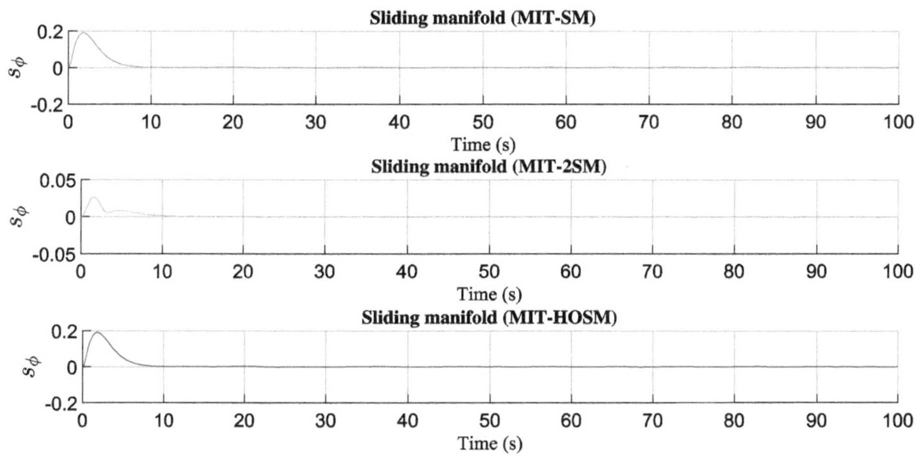

Figure 5.52 Response of the sliding manifold in the adaptive derivative gain for the roll angle (with disturbances).

Table 5.10 $\mathcal{L}_2$-norm of the errors and control efforts of the nonlinear controllers and the combinations of the nonlinear controllers.

Controller	Altitude [m]		Yaw [deg]		Roll [deg]	
	$\mathcal{L}_2[e_\theta]$	$\mathcal{L}_2[u_\theta]$	$\mathcal{L}_2[e_\psi]$	$\mathcal{L}_2[u_\psi]$	$\mathcal{L}_2[e_\phi]$	$\mathcal{L}_2[u_\phi]$
Nested saturation	6.3038	0.1950	7.1146	1.7905	2.7745	5.8278
Backstepping	1.6721	2.5093	5.3429	2.1544	2.8112	7.0331
SMC	1.9503	1.0105	0.6896	4.7391	0.2027	2.6379
Sat-SMC	5.6669	10.9178	7.0275	23.6873	2.4438	15.7763
Sat-2SMC	4.7044	1.8521	7.4052	8.8351	2.7181	10.2700
Sat-HOSM	2.8446	1.8521	7.1291	1.7791	3.4521	3.8505
Back-SMC	1.7718	19.9411	5.4711	25.0551	2.0546	27.9312
Back-2SMC	2.0452	6.4643	5.2187	1.6598	3.0817	22.3280
Back-HOSM	1.9503	5.3973	5.3429	2.1544	2.6157	6.6546
MIT	0.2301	0.0068	0.0820	0.0169	0.1964	0.1272
MIT-SMC	0.2136	0.0080	0.0416	0.0218	0.1892	0.1177
MIT-2SMC	0.2335	1.6718×10^{-4}	0.1861	0.2093	0.1768	0.3523
MIT-HOSM	0.2077	4.5471×10^{-4}	0.0359	0.7774	0.0780	2.6101

5.11 Concluding remarks

This chapter described the design of nonlinear controllers and the combination of some nonlinear control techniques using the aerodynamic model that defines the fixed-wing UAV shown in Chapter 3. The obtained simulations results were shown. Error analysis with the vectorial norms revealed what control law presented the most significant error. The same norm was used to obtain the effort applied by every controller to achieve the desired value.

It is worth mentioning that said controllers and analyses of error and effort were realized for the three fundamental movements of a fixed-wing UAV: altitude, yaw angle, and roll angle.

State observers

6

This chapter presents the design of linear and nonlinear state observers, which are applied to the altitude movement and the yaw and roll angles. The linear observer is the complementary filter presented in Section 6.2. The nonlinear observer by sliding mode is shown in Section 6.3, NESO is developed in Section 6.4, and Section 6.5 is devoted to the backstepping observer.

6.1 Applications and concepts of state observers in control theory

Control systems regulate various machines, products, and manufacturing processes. Some control variables are motion, temperature, liquid flow, voltage, current, torque, etc. A common assumption is the availability of measurement by digital or analogical sensors, but this is not always true. Any digital or analogical sensor has deficiencies that adversely affect a control system. Some sensors' problems are the following: the wiring reduces its reliability, hostile environments could make signal acquisition difficult or impossible, measurements suffer from stochastic noise, low sampling rates, and high costs.

Observers are algorithms that combine the reading provided by one or many sensors with partial or overall knowledge of the control system behavior. In some cases, the observer's signal could be more accurate, have less computational cost, and be more reliable than the sensor's signal. In another way, the observers could replace the sensors to obtain a more reliable signal. The observer output is named observer signal.

It is impossible to reject situations like parametric uncertainties, external perturbations by nonmodeled dynamics, minor measurement errors, modeling linearization, etc. Thus, a robust observer design is advisable to lessen the problems caused by the situations mentioned above. The robust observer has two aims: to improve the measurement's precision and to observe all or a few perturbations. Therefore, the problem of designing a robust state observer for dynamic systems with significant perturbations is an engaging research topic.

Below we list some reports on state observers for nonlinear dynamic systems that have been developed in recent years. The robust state observer presented in [70] considers nonlinear bounded uncertainties, and the observer is designed by the system variable structure method. However, the asymptotic convergence between observer state and real state (i.e., observation error) is not guaranteed. Years later, an improvement ensured the exponential asymptotic convergence of the observation error [19]. In [25], the authors use the algebraic solution of Riccati's equation to develop a new robust state observer for dynamic systems with perturbations. The observer guarantees the observation error approaches zero if it is possible to satisfy some conditions. In [5] such a demonstration is shown.

Design of Control Laws and State Observers for Fixed-Wing UAVs. https://doi.org/10.1016/B978-0-32-395405-1.00015-7

Another typical problem is time delays among sampling because it is an instability source. The time delays are found in chemical processes, hydraulic systems, systems with bearings, and the economy. Hence, the time delay presence is an important study topic. Some developments to attend time delays are described in [16–49]. We are going to discuss the research on state observers for linear systems. In [21], the authors present an observer for time delay and sufficient conditions obtained from linear matrix inequalities, and the reference [22] shows the same case for discrete time.

A reduced-order observer for neutral linear systems presented in [57] proves stability despite time delay. In [58] a design for observers with unknown inputs is offered. The methodology presented in [16] solves the variable delay time case by an SMO. However, they assume that the time delay and time delay limits are known. Based on such knowledge, the authors propose several kinds of state observers. Most of the time, delay reports require linear states or normal bounded linear states in the literature reviewed (see [35], [36]).

It should be noted that the superior limit of the normal vector uncertainty is known. Among control systems, in applications it is difficult to know such limits; often, the limits are unknown or partially known. Knowing the superior limit, a robust state observer can be constructed.

For that reason, the suggested is an adaptive scheme giving rise to adaptive robust state observers. An example is [34], where convergence is proved experimentally but not theoretically. In [12] an adaptive robust state observer is proposed for nonlinear dynamic systems with perturbed states and asymptotic convergence to zero is demonstrated, despite significant uncertainties and time delays.

6.2 Complementary filters

The complementary filter is a filter written in Laplace variables. It refers to the use of two or more transient functions that functions are complementary between them. Moreover, if the data sensor is the input of a function $G(s)$, then the other data sensor is the input of $G(s)^{-1}$, and the result of all transient functions is the identity matrix. Fig. 4.1 depicts an example of a complementary filter for a system with two inputs and one output.

The first input has a high-frequency noise, and the second input a low-frequency noise, for which a low-pass filter and a high-pass filter must necessarily be applied, respectively. If both filters function complementarily, the output filter is the variable signal without noise. Assume that the flight data such as velocity, acceleration, and the trajectory traveled are available. Let be $\hat{\dot{x}}$ the estimated velocity defined as follows:

$$\hat{\dot{x}} = G(s)x_d + (1 - G(s))x_1, \tag{6.1}$$

where x_D is the position time derivative, x_I is the acceleration integral, and $G(s)$ is the low-pass filter

$$G(s) = \frac{\alpha}{s + \alpha} \tag{6.2}$$

with the complementary high-pass filter

$$1 - G(s) = \frac{s}{s+\alpha}. \tag{6.3}$$

6.3 Sliding mode observers

Soviets have generated most of the sliding mode literature for stabilizing some linear systems. The sliding mode controls show exceptional hardiness and properties for trajectory tracking in nonlinear systems; this method guarantees hardiness in a range of system uncertainties and perturbations.

Although sliding modes have excellent properties in theory against parametric uncertainties, this method presents two drawbacks that limit its implementation. The first one is a high computational cost, and the second one is chattering. Chattering is not desirable in practice because it could excite nonmodeled high-frequency dynamics, and control costs increase. The nonmodeled high-frequency dynamics could be structural resonance, actuator delays, or errors in sensor acquisitions.

The sliding mode observers offer a solution with similar properties as sliding mode control. As for drawbacks, the chattering has numeric implications only.

6.3.1 *Sliding surface*

This subsection is a brief revision of the sliding modes theory and discusses the potential advantages of discontinuous control laws. Assume the following dynamic system:

$$x^{(n)}(t) = f(x;t) + b(x;t)u(t) + d(t), \tag{6.4}$$

where $u(t)$ is a scalar control input, x is a scalar output, $\mathbf{x} = [x_1, x_2, x_3, \cdot, x_{n-1}]^\top$ are the states, and $d(t)$ is a perturbation. The function $f(\mathbf{x}, t)$ is known partially; the unknown terms are represented by $|\Delta f|$ and upper bounded by a known function of $\mathbf{x}$ and t. Similarly, gains are represented by $b(x;t)$ with sign and bounds known; also, the control gains must be a continuous function of x and t. Both $f(\mathbf{x};t)$ and $b(\mathbf{x};t)$ are supposed to be continuous on a domain of $\mathbf{x}$. The only knowledge about $d(t)$ is the superior bound for its absolute value. We aim to get the value of state $\mathbf{x}$ to achieve a specific state $\mathbf{x}_d = \left[x_d, \dot{x}_d, \cdot, \mathbf{x}_d^{(n-1)}\right]^\top$ although $f(\mathbf{x};t)$ and $b\mathbf{x}; y$ have uncertainties and perturbation $d(t)$ is present. Assume that $\mathbf{x}_d = \mathbf{x}$ from $t = 0$ with a finite control law u, so

$$\tilde{\mathbf{x}} |_{t=0} = \mathbf{0}, \tag{6.5}$$

where $\tilde{\mathbf{x}} := \mathbf{x} - \mathbf{x}_d = \left[\tilde{x}, \dot{\tilde{x}}, \cdots, \tilde{x}^{(n-1)}\right]^\top$ is the trajectory tracking error. This supposition will be analyzed later. Let $s(t)$ be a time-variant sliding surface in $\mathbf{R}^n$ with

$s(\mathbf{x};t) = 0$ and

$$s(\mathbf{x};t) := \left(\frac{d}{dt} + \lambda\right)^{n-1} \tilde{x}, \lambda > 0, \tag{6.6}$$

where λ is a positive constant. The initial condition (6.5) implies a perfect trajectory tracking; this means that $\mathbf{x} \equiv \mathbf{x}_d$ *is equivalent to hold in the sliding surface* $s(t)$ *for all* $t > 0$. Furthermore, $s \equiv 0$ represents a linear differential equation whose unique solution is $\tilde{x} \equiv 0$ for initial condition (6.5). We consider a control law u for (6.4) that is equivalent to $s(t)$,

$$\frac{1}{2}\frac{d}{dt}s^2(\mathbf{x};t) \leq -\eta|s|, \tag{6.7}$$

where η is a positive constant. Inequality (6.7) constrains the trajectories to point towards the surface $S(t)$ and it is named a *sliding condition.*

The purpose of Eqs. (6.6) and (6.7) is to achieve a right behavior of (6.6) and then select a control law u for (6.4) such that s^2 is a Lyapunov function for the closed-loop dynamic *despite the presence of inaccuracy in the model and disturbances*. Further, if (6.7) is satisfied, this guarantees that although Eq. (6.5) is not satisfied (i.e., if $\tilde{\mathbf{x}}|_{t=0} \neq \mathbf{0}$), $S(t)$ will be reached in a finite time less than $|s(\mathbf{x}(0);0)|/\eta$, while the definition (6.6) guarantees that $\tilde{\mathbf{x}} \to 0$ when $t \to \infty$. The control law that satisfies (6.7), however, has to be *discontinuous* across the sliding surface, which leads in practice to small oscillations called chattering.

The problem, similarly exploiting the sliding behavior in the design of *observers*, rather than controllers, is precisely that the total of states is not available to measure, and therefore a definition of sliding surface analogous to (6.6) is not suitable. Intuition can be developed to address this difficulty by simply considering second-order dynamics.

6.3.2 Shear effect and sliding patch

Let us consider the generation of slip behavior in a second-order system through input switching according to the value of a single state component, instead of a linear combination of both components, as in (6.6). The system

$$\dot{x}_1 = x_2, \tag{6.8}$$

$$\dot{x}_2 = -k_2\,\mathrm{sgn}(x_1), \tag{6.9}$$

$$\dot{x}_1 = x_2 - k_2\,\mathrm{sgn}(x_1), \tag{6.10}$$

$$\dot{x}_2 = -k_2\,\mathrm{sgn}(x_1), \tag{6.11}$$

where k_2 is a positive constant and sgn(·) is the sign function, clearly shows no sliding behavior. Instead, let us consider the system

$$|x_2| \leq k_1, \tag{6.12}$$

which has been referred to as a *sliding patch*. Furthermore, we have the condition

$$\frac{d}{dt}(x_1)^2 < 0. \tag{6.13}$$

This is satisfied if the condition (6.13) holds, which defines the slip patch. The dynamics over the slip patch itself can be derived from the concept of Filippov's definition, which formalizes the following engineering intuition: the dynamics over the patch can only be a convex combination (i.e., an average) of the dynamics of each side of the discontinuity surface:

$$\dot{\hat{x}}_1 = \gamma(x_2 + k_1) + (1-\gamma)(x_2 - k_1), \tag{6.14}$$

$$\dot{\hat{x}}_2 = \gamma k_1 + (1-\gamma)(-k_1). \tag{6.15}$$

The value of γ, and therefore the resulting dynamics, is then formally determined by the invariance of the patch itself:

$$\dot{x}_1 = 0 \Rightarrow \dot{x}_2 = -(k_2/k_1)x_2. \tag{6.16}$$

Thus, x_2 decreases exponentially to 0 after reaching the sliding patch, with a time constant k_1/k_2. Furthermore, it can easily be shown that all trajectories start on the x_2-axis and look for the patch in as little time as $|x_2(t=0)|/(k_1k_2)$. Currently, slippage can be guaranteed from time $t = 0$ by making k_1 and k_2 variants in time, with

$$k_2/k_1 \geq a, \tag{6.17}$$

$$k_1 > |x_2(t=0)|e^{-at}, \tag{6.18}$$

where $a = a(t)$ is any positive definite function in time.

6.3.3 System damping

Now consider the following system:

$$\dot{x}_1 = -\alpha_1 x_1 + x_2 - k_1 \operatorname{sgn}(x_1), \tag{6.19}$$

$$\dot{x}_2 = -\alpha_2 x_1 - k_2 \operatorname{sgn}(x_1). \tag{6.20}$$

By repeating the previous analysis, the slip condition is checked in the extended region:

$$x_1 = -\alpha_1 x_1 + x_2 - k_1 \operatorname{sgn}(x_1), \tag{6.21}$$

$$x_2 = -\alpha_2 x_1 - k_2 \operatorname{sgn}(x_1). \tag{6.22}$$

Then, the sum of the damping term at α_1 increases the region of direct attraction. Also, the value of α_2 only affects the phase capture but not the dynamics of the patch itself, which remains unchanged:

$$\dot{x}_2 = -(k_2/k_1)x_2. \tag{6.23}$$

From [59] for the system

$$\ddot{x} = f, \tag{6.24}$$

where f is a nonlinear function with state uncertainty x, an observer with a structure of the following form can be used:

$$\begin{aligned} \dot{\hat{x}}_1 &= -\alpha_1 \tilde{x}_1 + \hat{x}_2 - k_1 \operatorname{sgn}(\hat{x}_1), \\ \dot{\hat{x}}_2 &= -\alpha_2 \tilde{x}_1 + \hat{f} - k_2 \operatorname{sgn}(\hat{x}_1), \end{aligned} \tag{6.25}$$

where $\tilde{x}_1 = \hat{x}_1 - x_1$, the constants α_j are chosen as a Luenberger observer in order to place the poles of the linearized system at the desired positions, and k_1 and k_2 are positive constants.

6.4 Nonlinear extended state observer

In [37], an alternative method is proposed where f is unknown in the system. Now we have

$$\ddot{y} = f(y, \dot{y}, w) + b_0 u, \tag{6.26}$$

where f represents the dynamics of the plant and is normally nonlinear, w is an unknown input disturbance, u is the control signal, b_0 is assumed to be known, and y is the output measure. In equations of state, the system is defined as

$$\begin{cases} \dot{x}_1 = x_2, \\ \dot{x}_2 = x_3 + b_0 u, \\ \dot{x}_3 = h, \\ y = x_1, \end{cases} \tag{6.27}$$

where $f(y, \dot{y}, w)$ is treated as another state, x_3. From the previous system, both $f(y, \dot{y}, w)$ and its derivative $h = \dot{f}(y, \dot{y}, w)$ are assumed unknown. However, by making $f(y, \dot{y}, w)$ a state, it can now be estimated. For this in [37] the following nonlinear observer is proposed:

$$\begin{cases} \dot{z}_1 = z_2 + \beta_1 g_1(e), \\ \dot{z}_2 = z_3 + \beta_2 g_2(e) + b_0 u, \\ \dot{z}_3 = \beta_3 g_3(e), \end{cases} \tag{6.28}$$

where $e = y - z_1$, z_3 is the estimate of the uncertainty function $f(\cdot)$, and $g_i(\cdot)$ is a modified exponential gain function.

The function $g_i(\cdot)$ is defined as

$$g_i(e, \alpha_i, \delta) = \begin{cases} |e|^{\alpha_i} \operatorname{sgn}(e), & |e| > \delta, \\ \frac{e}{\delta^{1-\alpha_i}}, & |e| \le \delta, \end{cases} \quad \delta > 0, \tag{6.29}$$

where α is chosen between 0 and 1, allowing g_i to have a high gain when the error is small, and δ is a small number used to limit the gain in the source neighborhood.

6.5 Backstepping observer

Consider the system:

$$\dot{x}_1 = x_2, \tag{6.30}$$
$$\dot{x}_2 = c_1 x_2 + c_2 u \tag{6.31}$$

and the full-order Luenberger observer

$$\dot{\hat{x}}_1 = \hat{x}_2 + k_1 \left(x_1 - \hat{x}_1\right), \tag{6.32}$$
$$\dot{\hat{x}}_2 = c_1 \hat{x}_2 + c_2 u + k_2 \left(x_1 - \hat{x}_1\right) \tag{6.33}$$

with

$$\hat{y} = \hat{x}_1. \tag{6.34}$$

The tracking error is introduced as

$$z_d = \begin{bmatrix} y_d \\ \dot{y}_d \end{bmatrix}, \quad \hat{z} = \begin{bmatrix} \hat{y} \\ \dot{\hat{y}} \end{bmatrix}, \quad z = z_d - \hat{z}$$

and the estimation error as

$$\tilde{x} = x - \hat{x},$$

where y_d is the twice differentiable desired trajectory. Then

$$\hat{y} = \hat{x}_1, \tag{6.35}$$
$$\dot{\hat{y}} = \dot{\hat{x}} = \hat{x}_2 + k_1 \left(x_1 - \hat{x}_1\right), \tag{6.36}$$
$$\ddot{\hat{y}} = \dot{\hat{x}}_2 + k_1 \left(\dot{x}_1 - \dot{\hat{x}}_1\right), \tag{6.37}$$
$$\ddot{\hat{y}} = c_1 \hat{x}_2 + c_2 u + k_2 \left(x_1 - \hat{x}_1\right) + k_1 \left(x_2 - \hat{x}_2 - k_1 \left(x_1 - \hat{x}_1\right)\right), \tag{6.38}$$
$$\ddot{\hat{y}} = c_1 \hat{x}_2 + c_2 u + k_2 \tilde{x}_1 + k_1 \tilde{x}_2 - k_1^2 \tilde{x}_1, \tag{6.39}$$
$$\ddot{\hat{y}} = c_1 \hat{x}_2 + c_2 u + \left(k_2 - k_1^2\right) \tilde{x}_1 + k_1 \tilde{x}_2. \tag{6.40}$$

The new system is

$$z_1 = y_d - \hat{y}, \tag{6.41}$$
$$z_1 = z_d - \hat{z}_1, \tag{6.42}$$

$$\dot{z}_1 = \dot{z}_d - \dot{\hat{z}}_1, \tag{6.43}$$

$$\dot{z}_1 = \dot{z}_d - \hat{z}_2, \tag{6.44}$$

$$z_2 = \dot{y} - \dot{\hat{y}}, \tag{6.45}$$

$$\dot{z}_2 = \ddot{y}_d - \ddot{\hat{y}}, \tag{6.46}$$

$$\dot{z}_2 = \ddot{z}_d - c_1 x_2 - c_2 u - \left(k_2 - k_1^2\right)\tilde{x}_1 - k_1 \tilde{x}_2, \tag{6.47}$$

$$\dot{z}_1 = \dot{z}_d - \hat{z}_2, \tag{6.48}$$

$$\dot{z}_2 = \ddot{z}_d - c_1 x_2 - c_2 u - \left(k_2 - k_1^2\right)\tilde{x}_1 - k_1 \tilde{x}_2. \tag{6.49}$$

The first storage function using the backstepping technique is defined,

$$V_1(z) = \frac{1}{2} z_1^2, \tag{6.50}$$

whose derivative is

$$\dot{V}_1(z) = z_1 \dot{z}_1 = z_1 \left(\dot{z}_d - \hat{z}_2\right). \tag{6.51}$$

Taking $\hat{z}_2$ as a virtual control input, it can be proposed that

$$\hat{z}_2 = b_1 z_1 + \dot{z}_d. \tag{6.52}$$

When substituting this in $\dot{V}_1(z)$, we have

$$\dot{V}_1(z) = z_1 \left(\dot{z}_d - (b_1 z_1 + \dot{z}_d)\right), \tag{6.53}$$

$$\dot{V}_1(z) = z_1 \left(\dot{z}_d - b_1 z_1 - \dot{z}_d\right), \tag{6.54}$$

$$\dot{V}_1(z) = -b_1 z_1^2. \tag{6.55}$$

But $\hat{z}_2$ is a state, so it is proposed that

$$\hat{z}_{2_u} = b_1 z_1 + \dot{z}_d = \alpha(z). \tag{6.56}$$

Therefore, consider a new coordinate system:

$$\zeta = \begin{bmatrix} \zeta_1 \\ \zeta_2 \end{bmatrix} = \begin{bmatrix} z_1 \\ \alpha(z) - \hat{z}_2 \end{bmatrix}. \tag{6.57}$$

Then

$$\zeta_1 = z_1, \tag{6.58}$$

$$\dot{\zeta}_1 = \dot{z}_1, \tag{6.59}$$

$$= \dot{z}_d - \hat{z}_2 \tag{6.60}$$

$$
\begin{aligned}
&= \dot{z}_d - \alpha(z) + \zeta_2 && (6.61)\\
&= \dot{z}_d - b_1 z_1 - \dot{z}_d + \zeta_2 && (6.62)\\
\dot{\zeta}_1 &= \zeta_2 - b_1\zeta_1, && (6.63)
\end{aligned}
$$

$$
\begin{aligned}
\zeta_2 &= \alpha(z) - \hat{z}_2 && (6.64)\\
&= b_1\zeta_1 + \dot{z}_d - \hat{z}_2, && (6.65)\\
\dot{\zeta}_2 &= b_1\dot{\zeta}_1 + \ddot{z}_d - \dot{\hat{z}}_2 && (6.66)\\
&= b_1(\zeta_2 - b_1\zeta_1) + \ddot{z}_d - c_1\hat{x}_2 - c_2 u - \left(k_2 - k_1^2\right)\tilde{x}_1 - k_1\tilde{x}_2. && (6.67)
\end{aligned}
$$

Now, we proceed with the second stage of backstepping. For this, we define the second storage function as

$$
V_2(\zeta) = \frac{1}{2}\zeta_1^2 + \frac{1}{2}\zeta_2^2 \tag{6.68}
$$

and the derivative is

$$
\begin{aligned}
\dot{V}_2(\zeta) &= \zeta_1\dot{\zeta}_1 + \zeta_2\dot{\zeta}_2 && (6.69)\\
&= \zeta_1(\zeta_2 - b_1\zeta_1)\,\zeta_2\big(b_1(\zeta_2 - b_1\zeta_1) + \ddot{z}_d - c_1\hat{x}_2 - c_2 u && (6.70)\\
&\qquad - \left(k_2 - k_1^2\right)\tilde{x}_1 - k_1\tilde{x}_2\big)\\
&= -b_1\zeta_1^2 + \zeta_2\big(\zeta_1 + b_1(\zeta_2 - b_1\zeta_1) + \ddot{z}_d - c_1\hat{x}_2 - c_2 u && (6.71)\\
&\qquad - \left(k_2 - k_1^2\right)\tilde{x}_1 - k_1\tilde{x}_2\big).
\end{aligned}
$$

Since in this stage the real input of the system is found, then it is proposed that

$$
c_2 u = b_2\zeta_2 + \zeta_1 + b_1(\zeta_2 - b_1\zeta_1) + \ddot{z}_d - c_1\hat{x}_2 - \left(k_2 - k_1^2\right)\tilde{x}_1, \tag{6.72}
$$

resulting in

$$
\begin{aligned}
\dot{V}_2(\zeta) &= -b_1\zeta_1^2 + \zeta_2\big(\zeta_1 + b_1(\zeta_2 - b_1\zeta_1) + \ddot{z}_d - c_1\hat{x}_2 && (6.73)\\
&\quad - \left(b_2\zeta_2 + \zeta_1 + b_1(\zeta_2 - b_1\zeta_1) + \ddot{z}_d - c_1\hat{x}_2 - \left(k_2 - k_1^2\right)\tilde{x}_1\right)\\
&\quad - \left(k_2 - k_1^2\right)\tilde{x}_1 - k_1\tilde{x}_2\big),\\
\dot{V}_2(\zeta) &= -b_1\zeta_1^2 - b_2\zeta_2^2 - k_1\zeta_2\tilde{x}_2. && (6.74)
\end{aligned}
$$

As indicated in [55], we will search for a matrix $P_0 = P_0^T > 0$ and a matrix $Q_0 = Q_0^T > 0$ that satisfy the Lyapunov equation

$$
(A - KC)^T P_0 + P_0(A - KC) = -Q_0, \tag{6.75}
$$

with which the estimation error is exponentially stable, with the following properties of the Lyapunov function:

$$\begin{aligned} &V_0(\tilde{x}) = \tilde{x}^T P_0 \tilde{x} > 0, \quad \|\tilde{x}\| \neq 0, \\ &\frac{d}{dt} V_0 = -\tilde{x}^T Q_0 \tilde{x} < 0, \quad \|\tilde{x}\| \neq 0. \end{aligned} \tag{6.76}$$

Then, it is proposed that

$$Q_0 = \begin{bmatrix} 1 & 0 \\ 0 & 1 \end{bmatrix} \tag{6.77}$$

and then the matrix P_0 will be deduced according to Eq. (6.75). It is known that

$$A = \begin{bmatrix} 0 & 1 \\ 0 & g_1 \end{bmatrix}, \quad K = \begin{bmatrix} k_1 \\ k_2 \end{bmatrix}, \quad C = \begin{bmatrix} 1 & 0 \end{bmatrix}.$$

The above is substituted in Eq. (6.75) so that

$$\begin{aligned} &\left(\begin{bmatrix} 0 & 1 \\ 0 & g_1 \end{bmatrix} - \begin{bmatrix} k_1 \\ k_2 \end{bmatrix} \begin{bmatrix} 1 & 0 \end{bmatrix} \right)^T \begin{bmatrix} P_{11} & P_{12} \\ P_{21} & P_{22} \end{bmatrix} \\ &+ \begin{bmatrix} P_{11} & P_{12} \\ P_{21} & P_{22} \end{bmatrix} \left(\begin{bmatrix} 0 & 1 \\ 0 & g_1 \end{bmatrix} - \begin{bmatrix} k_1 \\ k_2 \end{bmatrix} \begin{bmatrix} 1 & 0 \end{bmatrix} \right) = - \begin{bmatrix} 1 & 0 \\ 0 & 1 \end{bmatrix}, \end{aligned} \tag{6.78}$$

$$\begin{aligned} &\left(\begin{bmatrix} 0 & 1 \\ 0 & g_1 \end{bmatrix} - \begin{bmatrix} k_1 & 0 \\ k_2 & 0 \end{bmatrix} \right)^T \begin{bmatrix} P_{11} & P_{12} \\ P_{21} & P_{22} \end{bmatrix} \\ &+ \begin{bmatrix} P_{11} & P_{12} \\ P_{21} & P_{22} \end{bmatrix} \left(\begin{bmatrix} 0 & 1 \\ 0 & g_1 \end{bmatrix} - \begin{bmatrix} k_1 & 0 \\ k_2 & 0 \end{bmatrix} \right) = - \begin{bmatrix} 1 & 0 \\ 0 & 1 \end{bmatrix}, \end{aligned} \tag{6.79}$$

$$\begin{aligned} &\begin{bmatrix} -k_1 & 1 \\ -k_2 & g_1 \end{bmatrix}^T \begin{bmatrix} P_{11} & P_{12} \\ P_{21} & P_{22} \end{bmatrix} + \begin{bmatrix} P_{11} & P_{12} \\ P_{21} & P_{22} \end{bmatrix} \begin{bmatrix} -k_1 & 1 \\ -k_2 & g_1 \end{bmatrix} \\ &\quad = - \begin{bmatrix} 1 & 0 \\ 0 & 1 \end{bmatrix}, \end{aligned} \tag{6.80}$$

$$\begin{aligned} &\begin{bmatrix} -k_1 & -k_2 \\ 1 & g_1 \end{bmatrix} \begin{bmatrix} P_{11} & P_{12} \\ P_{21} & P_{22} \end{bmatrix} + \begin{bmatrix} P_{11} & P_{12} \\ P_{21} & P_{22} \end{bmatrix} \begin{bmatrix} -k_1 & 1 \\ -k_2 & g_1 \end{bmatrix} \\ &= - \begin{bmatrix} 1 & 0 \\ 0 & 1 \end{bmatrix}, \end{aligned} \tag{6.81}$$

$$\begin{aligned} &\begin{bmatrix} -k_1 P_{11} - k_2 P_{21} & -k_1 P_{12} - k_2 P_{22} \\ P_{11} + g_1 P_{21} & P_{12} + g_1 P_{22} \end{bmatrix} + \begin{bmatrix} -k_1 P_{11} - k_2 P_{12} & P_{11} + g_1 P_{12} \\ -k_1 P_{21} - k_2 P_{22} & P_{21} + g_1 P_{22} \end{bmatrix} \\ &= - \begin{bmatrix} 1 & 0 \\ 0 & 1 \end{bmatrix}, \end{aligned} \tag{6.82}$$

$$\begin{bmatrix} -2k_1 P_{11} - k_2 (P_{12} + P_{21}) & P_{11} + P_{12}(g_1 - k_1) - k_2 P_{22} \\ P_{11} + P_{21}(g_1 - k_1) - k_2 P_{22} & 2g_1 P_{22} + P_{21} + P_{12} \end{bmatrix}$$

$$= -\begin{bmatrix} 1 & 0 \\ 0 & 1 \end{bmatrix}. \tag{6.83}$$

Then

$$-2k_1 P_{11} - k_2 (P_{12} + P_{21}) = -1, \tag{6.84}$$

$$P_{11} + P_{12} (g_1 - k_1) - k_2 P_{22} = 0, \tag{6.85}$$

$$P_{11} + P_{21} (g_1 - k_1) - k_2 P_{22} = 0, \tag{6.86}$$

$$2g_1 P_{22} + P_{21} + P_{12} = -1. \tag{6.87}$$

Solving P_{11} of (6.85),

$$P_{11} = k_2 P_{22} - P_{12} (g_1 - k_1), \tag{6.88}$$

and by substituting (6.88) in (6.86), we have

$$k_2 P_{22} - P_{12} (g_1 - k_1) + P_{12} (g_1 - k_1) - k_2 P_{22} = 0,$$

$$P_{21} (g_1 - k_1) = P_{12} (g_1 - k_1),$$

$$P_{21} = P_{12}. \tag{6.89}$$

After substituting (6.89) in (6.87),

$$2g_1 P_{22} + P_{12} + P_{12} = -1,$$

$$2g_1 P_{22} = -1 - 2P_{12},$$

$$P_{22} = -\left(\frac{1 + 2p_{12}}{2g_1}\right). \tag{6.90}$$

On the other hand, by substituting (6.90) in (6.88),

$$P_{11} = -k_2 \left(\frac{1 + 2P_{12}}{2g_1}\right) - P_{12} (g_1 - k_1),$$

$$P_{11} = -\frac{k_2}{2g_1} - \frac{2k_2 P_{12}}{2g_1} - P_{12} (g_1 - k_1),$$

$$P_{11} = -\frac{k_2}{2g_1} - \frac{k_2}{g_1} - P_{12} (g_1 - k_1),$$

$$P_{11} = -\frac{k_2}{2g_1} - P_{12} \left(\frac{k_2}{g_1} + (g_1 - k_1)\right). \tag{6.91}$$

Now, substituting (6.91) in (6.84),

$$2k_1 \left(\frac{k_2}{2g_1} + P_{12} \left(\frac{k_2}{g_1} + (g_1 - k_1)\right)\right) - 2k_2 P_{12} = -1,$$

$$\frac{k_1 k_2}{g_1} + 2P_{12} \left(\frac{k_1 k_2}{g_1} + k_1 (g_1 - k_1)\right) - 2k_2 P_{12} = -1,$$

$$
\begin{aligned}
2P_{12}\left(\frac{k_1k_2}{g_1}+k_1(g_1-k_1)-k_2\right) &= -1-\frac{k_1k_2}{g_1},\\
2P_{12}\left(\frac{k_1k_2+k_1(g_1-k_1)g_1-k_2g_1}{g_1}\right) &= \frac{-g_1-k_1g_2}{g_1},\\
2P_{12}\left(k_1k_2+k_1(g_1-k_1)g_1-k_1g_1\right) &= -g_1-k_1k_2,\\
P_{21} = \frac{-g_1-k_1k_2}{2\left(k_1k_2+k_1(g_1-k_1)g_1-k_2g_1\right)}&,\\
P_{12} = P_{21} = \frac{g_1+k_1k_2}{2(g_1-k_1)(k_2-g_1k_1)}&,
\end{aligned}
$$

and replacing P_{12} in (6.90),

$$
\begin{aligned}
P_{22} &= -\frac{1}{2g_1}-\frac{1}{g_1}P_{12}\\
&= -\frac{1}{2g_1}-\frac{1}{g_1}\left(\frac{g_1+k_1k_2}{2(g_1-k_1)(k_2-g_1k_1)}\right)\\
&= -\frac{1}{2g_1}\left(1+\frac{g_1+k_1k_2}{(g_1-k_1)(k_2-g_1k_1)}\right)\\
&= -\frac{1}{2g_1}\left(\frac{(g_1-k_1)(k_2-g_1k_1)}{(g_1-k_1)(k_2-g_1k_1)}+\frac{g_1+k_1k_2}{(g_1-k_1)(k_2-g_1k_1)}\right)\\
&= -\frac{1}{2g_1}\left(\frac{-k_1k_2-(g_1-k_1)g_1k_1+g_1k_2}{(g_1-k_1)(k_2-g_1k_1)}+\frac{g_1+k_1k_2}{(g_1-k_1)(k_2-g_1k_1)}\right)\\
&= -\frac{1}{2g_1}\left(\frac{-k_1k_2-(g_1-k_1)g_1k_1+g_1k_2+g_1+k_1k_2}{(g_1-k_1)(k_2-g_1k_1)}\right)\\
&= -\frac{1}{2g_1}\left(\frac{g_1\left(-(g_1-k_1)k_1+k_2+1\right)}{(g_1-k_1)(k_2-g_1k_1)}\right)\\
&= \frac{(g_1-k_1)k_1-k_2-1}{2(g_1-k_1)(k_2-g_1k_1)}.
\end{aligned}
$$

Next, P_{12} and P_{22} are substituted in (6.88):

$$
\begin{aligned}
P_{11} &= k_2\left(\frac{(g_1-k_1)k_1-k_2-1}{2(g_1-k_1)(k_2-g_1k_1)}\right)-\left(\frac{g_1+k_1k_2}{2(g_1-k_1)(k_2-g_1k_1)}\right)(g_1-k_1),\\
P_{11} &= \frac{g_1k_1k_2-k_1^2k_2-k_2^2-k_2-g_1^2+g_1k_1-g_1k_1k_2+k_1^2k_2}{2(g_1-k_1)(k_2-g_1k_1)},\\
P_{11} &= \frac{-k_2^2-k_2+g_1k_1-g_1^2}{2(g_1-k_1)(k_2-g_1k_1)},
\end{aligned}
$$

so the matrix P_0 is

$$
P_0 = \begin{bmatrix} \frac{-k_2^2-k_2+g_1k_1-g_1^2}{2(g_1-k_1)(k_2-g_1k_1)} & \frac{g_1+k_1k_2}{2(g_1-k_1)(k_2-g_1k_1)} \\ \frac{g_1+k_1k_2}{2(g_1-k_1)(k_2-g_1k_1)} & \frac{(g_1-k_1)k_1-k_2-1}{2(g_1-k_1)(k_2-g_1k_1)} \end{bmatrix}, \tag{6.92}
$$

so that if $P_0 > 0$, then $P_{11} > 0$, and if $|P_0| > 0$, then

$$P_{11} = \frac{-k_2^2 - k_2 + g_1 k_1 - g_1^2}{2(g_1 - k_1)(k_2 - g_1 k_1)} > 0. \tag{6.93}$$

Then P_{11} will be positive if it is considered that $k_1, k_2 > 0$ and that $g_1 < 0$.

Now for $|P_0| > 0$,

$$|P_0| = \frac{\left(-k_2^2 - k_2 + g_1 k_1 - g_1^2\right)\left((g_1 - k_1) k_1 - k_2 - 1\right) - (g_1 + k_1 k_2)^2}{4(g_1 - k_1)^2 (k_2 - g_1 k_1)^2} > 0. \tag{6.94}$$

It is easy to notice that the denominator is positive, so it remains to prove that

$$\begin{aligned}
\left(-k_2^2 - k_2 + g_1 k_1 - g_1^2\right)\left((g_1 - k_1) k_1 - k_2 - 1\right) - (g_1 + k_1 k_2)^2 &> 0, \\
\left(-k_2^2 - k_2 + g_1 k_1 - g_1^2\right)\left(g_1 k_1 - k_2 - k_2 - 1\right) - g_1^2 - 2 g_1 k_1 k_2 - k_1^2 k_2^2 &> 0, \\
-g_1 k_1 k_2^2 + k_1^2 k_2^2 + k_2^3 + k_2^2 - 4 g_1 k_1 k_2 + k_1^2 k_2 + k_2^2 + k_2 + g_1^2 k_1^2 & \\
-g_1 k_1^3 - g_1 k_1 - g_1^3 k_1 + g_1^2 k_1^2 + g_1^2 k_2 &> 0, \\
k_2 \left(k_2^2 + k_1^2 + 2k_2 - 2 g_1 k_1 + g_1^2 + 1\right) & \\
-g_1 k_1 \left(k_2^2 + k_1^2 + 2k_2 - 2 g_1 k_1 + g_1^2 + 1\right) &> 0, \\
(k_2 - g_1 k_1)\left(k_2^2 + k_1^2 + 2k_2 - 2 g_1 k_1 + g_1^2 + 1\right) &> 0.
\end{aligned}$$

The second term is positive due to the same considerations made for P_{11}. Then

$$\begin{aligned}
k_2 - g_1 k_1 &> 0, \\
k_2 &> g_1 k_1, \\
g_1 &< \frac{k_2}{k_1}.
\end{aligned}$$

It is concluded that for P_0 to be positive definite, $k_1 > 0$, $k_2 > 0$, $g_1 < 0$, and $g_1 < \frac{k2}{k1}$.

Now it will be verified that if $V(\tilde{x}) = \beta \tilde{x}^T P_0 \tilde{x} > 0$, then $\dot{V}(\tilde{x}) = -\beta \tilde{x}^T Q_0 \tilde{x}$, where β is a design constant:

$$V(\tilde{x}) = \beta \begin{bmatrix} \tilde{x}_1 & \tilde{x}_2 \end{bmatrix} \begin{bmatrix} \frac{-k_2^2 - k_2 + g_1 k_1 - g_1^2}{2(g_1 - k_1)(k_2 - g_1 k_1)} & \frac{g_1 + k_1 k_2}{2(g_1 - k_1)(k_2 - g_1 k_1)} \\ \frac{g_1 + k_1 k_2}{2(g_1 - k_1)(k_2 - g_1 k_1)} & \frac{(g_1 - k_1) k_1 - k_2 - 1}{2(g_1 - k_1)(k_2 - g_1 k_1)} \end{bmatrix} \begin{bmatrix} \tilde{x}_1 \\ \tilde{x}_2 \end{bmatrix},$$

$$V(\tilde{x}) = \beta \begin{bmatrix} \tilde{x}_1 & \tilde{x}_2 \end{bmatrix} \begin{bmatrix} \frac{(-k_2^2 - k_2 + g_1 k_1 - g_1^2)\tilde{x}_1}{2(g_1 - k_1)(k_2 - g_1 k_1)} & \frac{(g_1 + k_1 k_2)\tilde{x}_2}{2(g_1 - k_1)(k_2 - g_1 k_1)} \\ \frac{(g_1 + k_1 k_2)\tilde{x}_1}{2(g_1 - k_1)(k_2 - g_1 k_1)} & \frac{((g_1 - k_1) k_1 - k_2 - 1)\tilde{x}_2}{2(g_1 - k_1)(k_2 - g_1 k_1)} \end{bmatrix},$$

$$V(\tilde{x}) = \beta \left[\frac{\left(-k_2^2 - k_2 + g_1 k_1 - g_1^2\right)\tilde{x}_1^2}{2(g_1 - k_1)(k_2 - g_1 k_1)} + \frac{(g_1 + k_1 k_2)\tilde{x}_1 \tilde{x}_2}{2(g_1 - k_1)(k_2 - g_1 k_1)} \right.$$

$$+\frac{(g_1+k_1k_2)\tilde{x}_1\tilde{x}_2}{2(g_1-k_1)(k_2-g_1k_1)}+\frac{((g_1-k_1)k_1-k_2-1)\tilde{x}_2^2}{2(g_1-k_1)(k_2-g_1k_1)}\Bigg],$$

$$V(\tilde{x})=\beta\frac{\left(-k_2^2-k_2+g_1k_1-g_1^2\right)\tilde{x}_1^2}{2(g_1-k_1)(k_2-g_1k_1)}+\beta\frac{(g_1+k_1k_2)\tilde{x}_1\tilde{x}_2}{(g_1-k_1)(k_2-g_1k_1)}$$
$$+\beta\frac{((g_1-k_1)k_1-k_2-1)\tilde{x}_2^2}{2(g_1-k_1)(k_2-g_1k_1)}.$$

Now we calculate the derivative of $V(\tilde{x})$:

$$\dot{V}(\tilde{x})=\beta\frac{\left(-k_2^2-k_2+g_1k_1-g_1^2\right)\tilde{x}_1\dot{\tilde{x}}_1}{(g_1-k_1)(k_2-g_1k_1)}+\beta\frac{(g_1+k_1k_2)\left(\tilde{x}_1\dot{\tilde{x}}_2+\tilde{x}_2\dot{\tilde{x}}_1\right)}{(g_1-k_1)(k_2-g_1k_1)}$$
$$+\beta\frac{((g_1-k_1)k_1-k_2-1)\tilde{x}_2\dot{\tilde{x}}_2}{(g_1-k_1)(k_2-g_1k_1)}.$$

Recall that

$$\dot{\tilde{x}}_1=x_2-\hat{x}_2-k_1\tilde{x}_1=\tilde{x}_2-k_1\tilde{x}_1,$$
$$\dot{\tilde{x}}_2=g_1x_2+g_2u-g_1\hat{x}_2-g_2u-k_2\tilde{x}_1=g_1\tilde{x}_2-k_2\tilde{x}_1.$$

Substituting this in $\dot{V}(\tilde{x})$, we have

$$\dot{V}(\tilde{x})=\beta\frac{\left(-k_2^2-k_2+g_1k_1-g_1^2\right)\tilde{x}_1(\tilde{x}_2-k_1\tilde{x}_1)}{(g_1-k_1)(k_2-g_1k_1)}$$
$$+\beta\frac{(g_1+k_1k_2)(\tilde{x}_1(g_1\tilde{x}_2-k_2\tilde{x}_1)+\tilde{x}_2(\tilde{x}_2-k_1\tilde{x}_1))}{(g_1-k_1)(k_2-g_1k_1)}$$
$$+\beta\frac{((g_1-k_1)k_1-k_2-1)\tilde{x}_2(g_1\tilde{x}_2-k_2\tilde{x}_1)}{(g_1-k_1)(k_2-g_1k_1)}$$
$$=-\beta k_1\frac{\left(-k_2^2-k_2+g_1k_1-g_1^2\right)\tilde{x}_1^2}{(g_1-k_1)(k_2-g_1k_1)}$$
$$+\beta\frac{\left(-k_2^2-k_2+g_1k_1-g_1^2\right)\tilde{x}_1\tilde{x}_2}{(g_1-k_1)(k_2-g_1k_1)}$$
$$+\beta\frac{(g_1+k_1k_2)\left(g_1\tilde{x}_1\tilde{x}_2-k_2\tilde{x}_1^2+\tilde{x}_2^2-k_1\tilde{x}_1\tilde{x}_2\right)}{(g_1-k_1)(k_2-g_1k_1)}$$
$$+\beta g_1\frac{((g_1-k_1)k_1-k_2-1)\tilde{x}_2^2}{(g_1-k_1)(k_2-g_1k_1)}$$
$$-\beta k_2\frac{((g_1-k_1)k_1-k_2-1)\tilde{x}_1\tilde{x}_2}{(g_1-k_1)(k_2-g_1k_1)}$$
$$=\beta\tilde{x}_1^2\left(-k_1\frac{\left(-k_2^2-k_2+g_1k_1-g_1^2\right)}{(g_1-k_1)(k_2-g_1k_1)}-k_2\frac{(g_1+k_1k_2)}{(g_1-k_1)(k_2-g_1k_1)}\right)$$

$$+\beta\tilde{x}_1\tilde{x}_2\left(\frac{(-k_2^2-k_2+g_1k_1-g_1^2)}{(g_1-k_1)(k_2-g_1k_1)}+\frac{(g_1+k_1k_2)(g_1-k_1)}{(g_1-k_1)(k_2-g_1k_1)}\right.$$
$$\left.-k_2\frac{((g_1-k_1)k_1-k_2-1)}{(g_1-k_1)(k_2-g_1k_1)}\right)$$
$$+\beta\tilde{x}_2^2\left(\frac{(g_1+k_1k_2)}{(g_1-k_1)(k_2-g_1k_1)}+g_1\frac{((g_1-k_1)k_1-k_2-1)}{(g_1-k_1)(k_2-g_1k_1)}\right),$$

$$\dot{V}(\tilde{x})=\beta\tilde{x}_1^2\left(\frac{k_1k_2^2+k_1k_2-g_1k_1^2+g_1^2k_1-g_1k_2-k_1k_2^2}{(g_1-k_1)(k_2-g_1k_1)}\right)$$
$$+\beta\tilde{x}_1\tilde{x}_2\left(\frac{-k_2^2-k_2+g_1k_1-g_1^2+g_1^2-g_1k_1}{(g_1-k_1)(k_2-g_1k_1)}\right.$$
$$\left.+\frac{g_1k_1k_2-k_1^2k_2-g_1k_1k_2+k_1^2k_2+k_2^2+k_2}{(g_1-k_1)(k_2-g_1k_1)}\right)$$
$$+\beta\tilde{x}_2^2\left(\frac{g_1+k_1k_2+g_1^2k_1-g_1k_1^2-g_1k_2-g_1}{(g_1-k_1)(k_2-g_1k_1)}\right)$$
$$=\beta\tilde{x}_1^2\left(-\frac{(g_1-k_1)(k_2-g_1k_1)}{(g_1-k_1)(k_2-g_1k_1)}\right)$$
$$+\beta\tilde{x}_1\tilde{x}_2\left(\frac{0}{(g_1-k_1)(k_2-g_1k_1)}\right)$$
$$+\beta\tilde{x}_2^2\left(-\frac{(g_1-k_1)(k_2-g_1k_1)}{(g_1-k_1)(k_2-g_1k_1)}\right),$$
$$\dot{V}(\tilde{x})=-\beta\tilde{x}_1^2-\beta\tilde{x}_2^2.$$

The Lyapunov candidate function is then defined as

$$V_L(\zeta,\tilde{x})=V_2(\zeta)+\beta\tilde{x}^TP_0\tilde{x}\quad\text{with }\beta>0, \tag{6.95}$$

where its derivative is given by

$$\dot{V}_L(\zeta,\tilde{x})=-b_1\zeta_1^2-b_2\zeta_2^2-k_1\zeta_2\tilde{x}_2-\beta\tilde{x}_1^2-\beta\tilde{x}_2^2. \tag{6.96}$$

Rewriting this in matrix form, we have

$$\dot{V}_L(\zeta,\tilde{x})=-\begin{bmatrix}\zeta_1 & \zeta_2 & \tilde{x}_1 & \tilde{x}_2\end{bmatrix}\begin{bmatrix}b_1 & 0 & 0 & 0\\ 0 & b_2 & 0 & \frac{k_1}{2}\\ 0 & 0 & \beta & 0\\ 0 & \frac{k_1}{2} & 0 & \beta\end{bmatrix}\begin{bmatrix}\zeta_1\\ \zeta_2\\ \tilde{x}_1\\ \tilde{x}_2\end{bmatrix}. \tag{6.97}$$

If

$$M_V = \begin{bmatrix} b_1 & 0 & 0 & 0 \\ 0 & b_2 & 0 & \frac{k_1}{2} \\ 0 & 0 & \beta & 0 \\ 0 & \frac{k_1}{2} & 0 & \beta \end{bmatrix}, \tag{6.98}$$

then M_V must be positive definite ($M_V > 0$) and then $\dot{V}_L(\zeta, \tilde{x}) < 0$. By design it is known that $b_1, b_2, k_1, \beta > 0$, so that if $M_V > 0$, then $|M_V| > 0$; performing the calculation we have

$$|M_V| = b_1 \left(b_2\beta^2 + \frac{k_1}{2}\left(-\frac{k_1\beta}{2}\right)\right) > 0,$$

$$b_1 b_2 \beta^2 - \frac{k_1^2 \beta b_1}{4} > 0,$$

$$\beta > \frac{k_1^2}{4b_2}. \tag{6.99}$$

6.6 Simulation results of the control laws with observers

Following are simulation results from the Luenberger observer, SMO, and NESO. The simulations were carried out in a closed loop using two different control laws. The PD control law is the first to be considered in the simulations. Later, a backstepping control law is chosen to compare performance, at least in simulation.

6.6.1 *PD control law with observers*

To test and analyze the three observers' behavior, we use a simple PD controller. Since the three observers obtain the estimation of all the states in the system, it will be the estimated states that will feedback the control law PD, which is known to use the position and speed of the system for regulation. With the estimated states feeding back the PD control law, the observers' efficiency is checked.

In the case of the control law for pitch motion, it was decided to control the altitude of the UAV as in Chapter 4, so the feedback error is defined as $e_h = h_d - h$. This error denotes the difference between the desired altitude h_d and the measured altitude h. Since the pitch control is related to the pitch motion, $\theta_d = \arctan(e_h/l)$ is proposed, which is the angle pitch that depends on the altitude error. The control signal PD is given by

$$u = k_{pj}(e_j) + k_{vj}(\dot{e}_j), \tag{6.100}$$

where k_{pj} and k_{vj} are the control gains related to the control law PD [52], [42]. The parameter j denotes θ, ψ, or ϕ, depending on the error of the control law for each movement, for example, for altitude it is $e_\theta = \theta_d - \hat{\theta}$.

For the yaw and roll movements, we use the general structure of the PD control of Eq. (6.100), considering the error equations as $e_\psi = \psi_d - \hat{\psi}$ and $e_\phi = \phi_d - \hat{\phi}$, where ψ_d and ϕ_d denote the desired yaw and roll angles and $\hat{\psi}$ and $\hat{\phi}$ are obtained from the Luenberger, SMO, or NESO observers.

On the other hand, for the dynamics of the altitude control we have $\dot{e}_\theta = \dot{\theta}_d - \dot{\hat{\theta}} = \dot{\theta}_d - \hat{q}$, where $\hat{q}$ is obtained from the Luenberger, SMO, or NESO observers. Similar arguments apply for yaw and roll movements.

6.6.2 Backstepping control law with observers

Consider the most general form of the backstepping control system reported in [41]:

$$\dot{x}_1 = f_b(x_1) + g_b(x_1)x_2, \tag{6.101}$$

$$\dot{x}_2 = f_a(x_1, x_2) + g_a(x_1, x_2)u, \tag{6.102}$$

where f_b, g_b, f_a, and g_a are known parameters and $f_b : D \to \mathbb{R}$ and $g_b : D \to \mathbb{R}$ are defined in the domain $D \subset \mathbb{R}^2$ containing $x_1 = 0$ and $f_b(0) = 0$. If $g_a(x_1, x_2) \neq 0$ on the domain of interest, the input is

$$u = \frac{1}{g_a(x_1, x_2)}(u_a - f_a(x_1, x_2)), \tag{6.103}$$

where Eq. (6.102) reduces to the integrator $\dot{x}_2 = u_a$. If a state feedback control law stabilizes at $\xi(\eta)$ and there exists a Lyapunov function $V(\eta)$ such that the conditions of the backstepping lemmas in [41] are satisfied for (6.101), then (6.103) is a state feedback control law to stabilize the system (6.101)–(6.102).

According to the dynamics of the system of observers seen in this book, which are of second order, it was established for the uncoupled model of the airplane that $f_b = 0$ and $g_b = 1$, which especially satisfies $f_b(0) = 0$.

6.6.3 Roll motion simulations with PD control law with observers

It is important to mention that both in the case of the roll and yaw movements, the simulations start from an error of 10 degrees since the desired angle for the two movements is zero degrees. In general, the simulations last 10 seconds.

Fig. 6.1 shows the simulation results for the roll motion of three state observers, Luenberger, SMO, and NESO; the three observers use a PD control law.

Fig. 6.2 shows the simulation results for the roll angular velocity of the three state observers.

Figs. 6.3 and 6.4 show the angular velocity and angle errors, respectively, of the three state observers with the PD control law. The simulation result for the extended state of NESO is presented in Fig. 6.5.

Fig. 6.6 shows the PD control signal of the simulation for the roll motion.

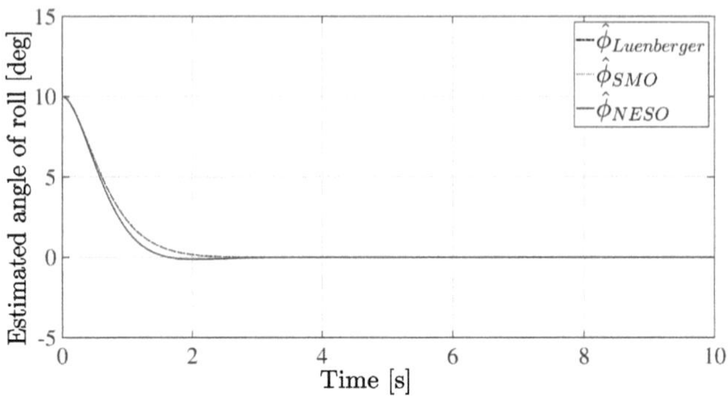

Figure 6.1 Roll angle estimation for the three state observers.

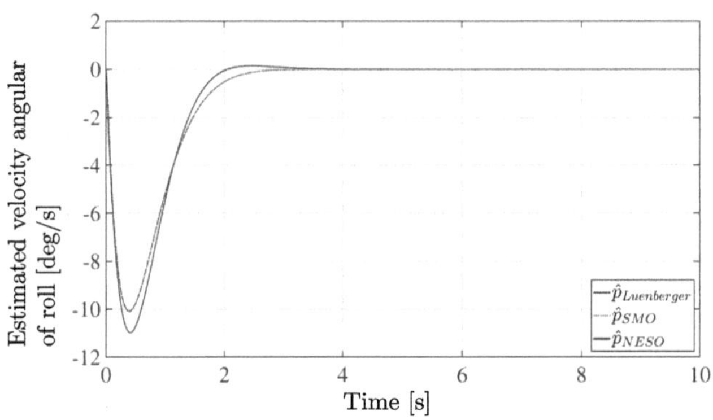

Figure 6.2 Roll angular velocity estimate for the three state observers.

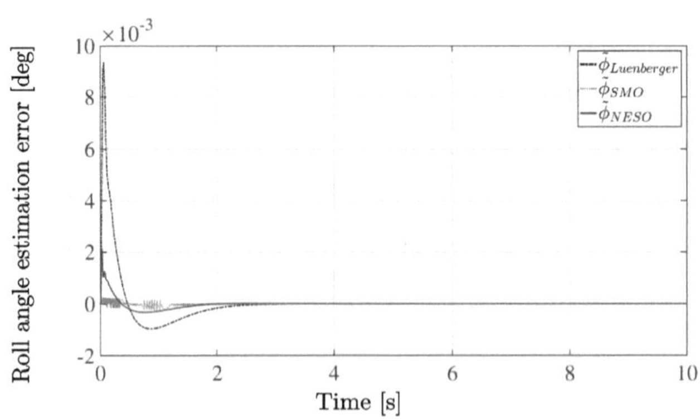

Figure 6.3 Roll angle estimation error for the three state observers.

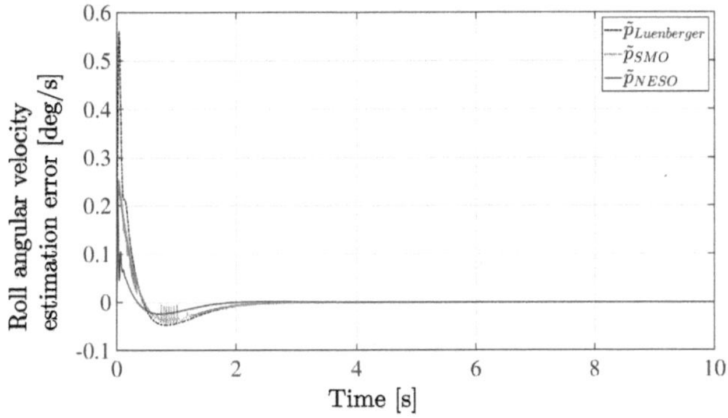

Figure 6.4 Roll angular velocity estimation error for the three state observers.

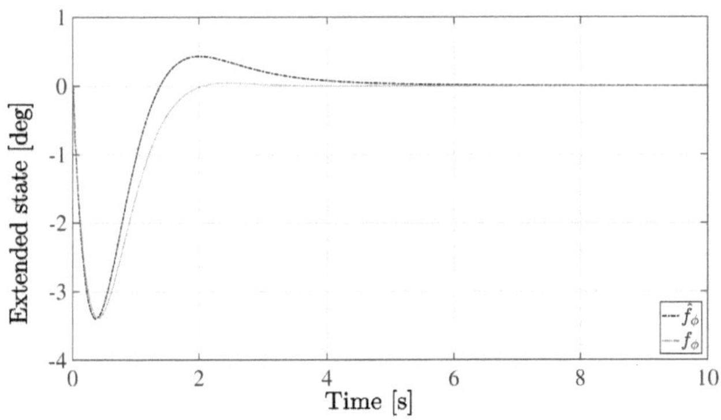

Figure 6.5 Extended state of NESO.

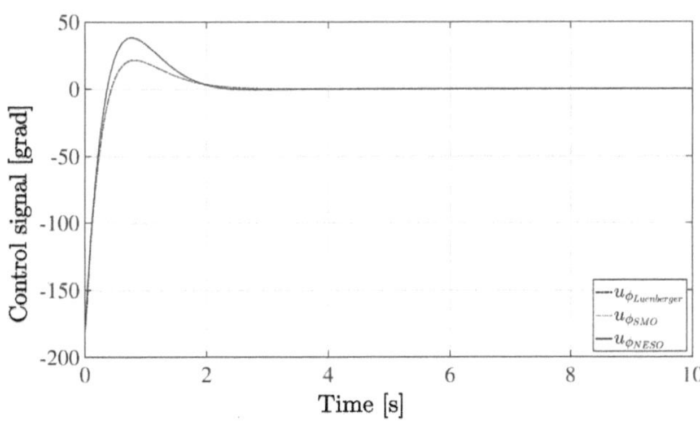

Figure 6.6 PD control signal for roll motion.

6.6.4 Yaw motion simulations with PD control law with observers

Shown below are the simulation results for the yaw motion with the PD control law. Fig. 6.7 shows the results of the three state observers for the yaw angle.

In Fig. 6.8 the results of the three observers for yaw angular velocity are presented. Figs. 6.9 and 6.10 present the simulation results of the angular position and angular velocity errors, respectively. The simulation results for NESO for the yaw motion can be seen in Fig. 6.11.

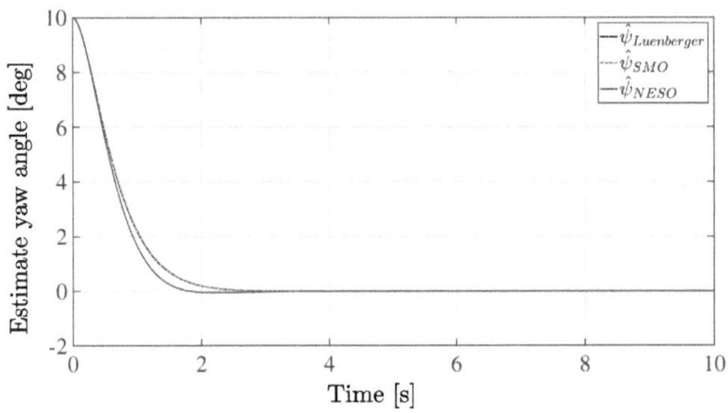

Figure 6.7 Yaw angle estimation for the three state observers.

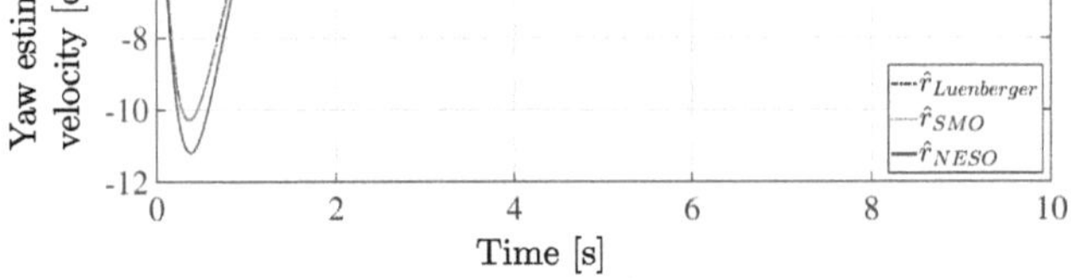

Figure 6.8 Yaw rate estimation for the three state observers.

The PD control signal for the yaw movement is shown in Fig. 6.12.

6.6.5 Altitude motion simulations with PD control law with observers

We present the simulation results for the altitude control fed back with the states estimated by the three observed ones presented in this work. Fig. 6.13 shows the simulation results for the pitch movement of the three observers.

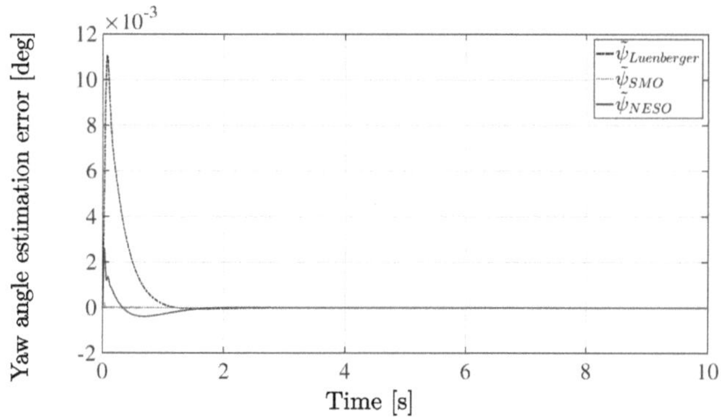

Figure 6.9 Yaw angle estimation error for the three state observers.

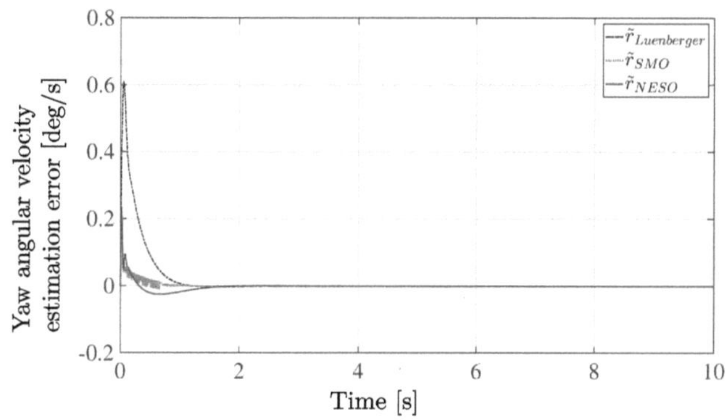

Figure 6.10 Yaw angular rate estimation error for the three state observers.

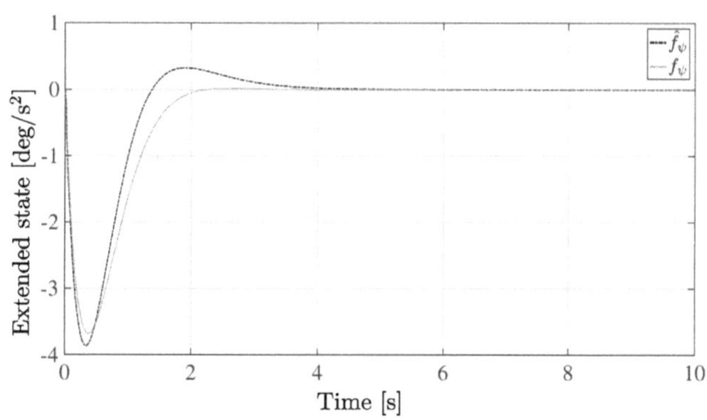

Figure 6.11 Extended state for yaw movement.

The simulation results for the pitch angular velocity of the three observers are presented in Fig. 6.14.

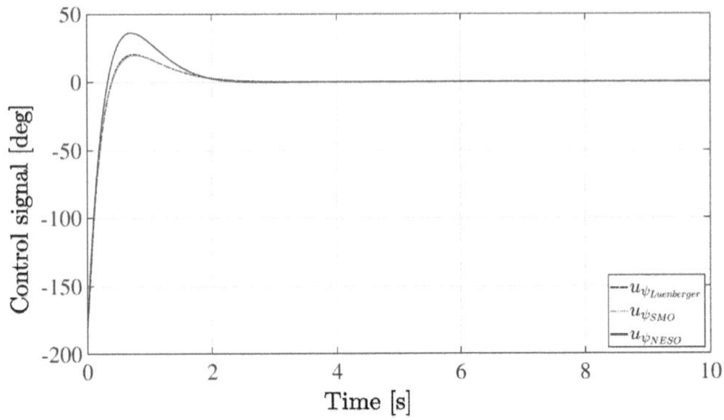

Figure 6.12 PD control signal for yaw movement.

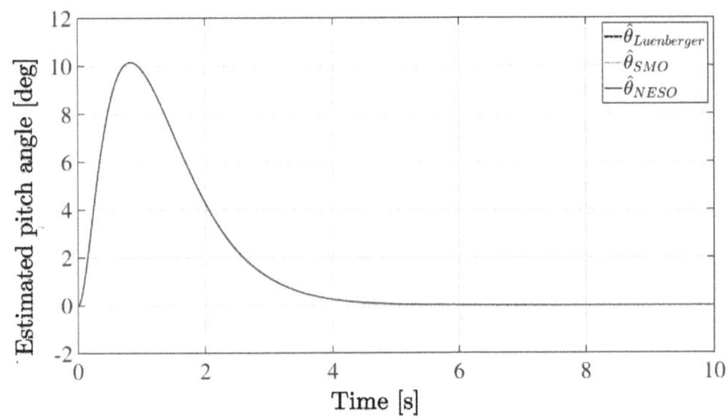

Figure 6.13 Pitch angle estimation for the three state observers.

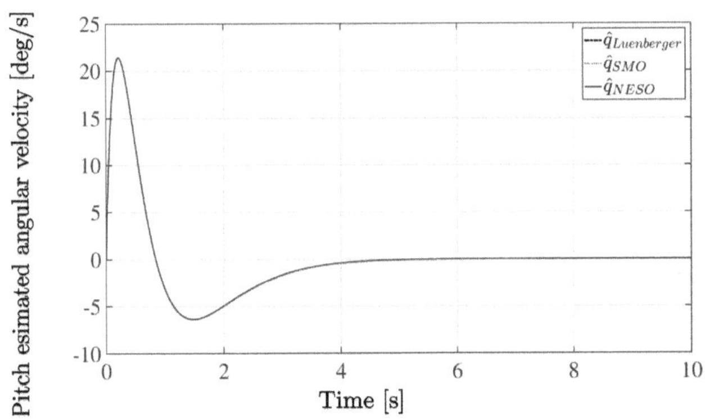

Figure 6.14 Pitch angular velocity estimate for the three state observers.

Figs. 6.15 and 6.16 show the simulation results of the errors of estimation of pitch angle and pitch angular velocity, respectively. The extended state of NESO applied to altitude movement is shown in Fig. 6.17.

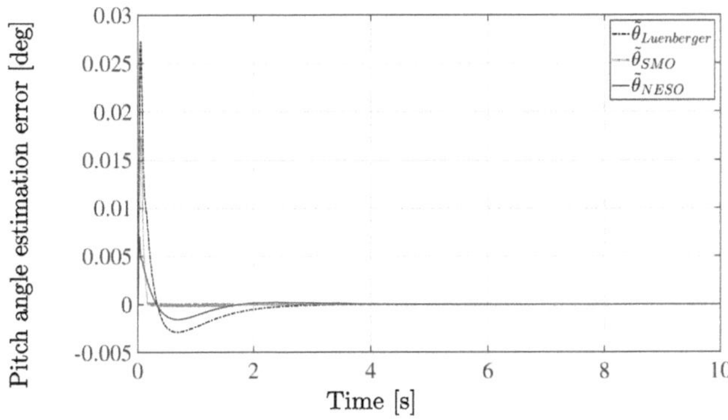

Figure 6.15 Pitch angle estimation error for the three state observers.

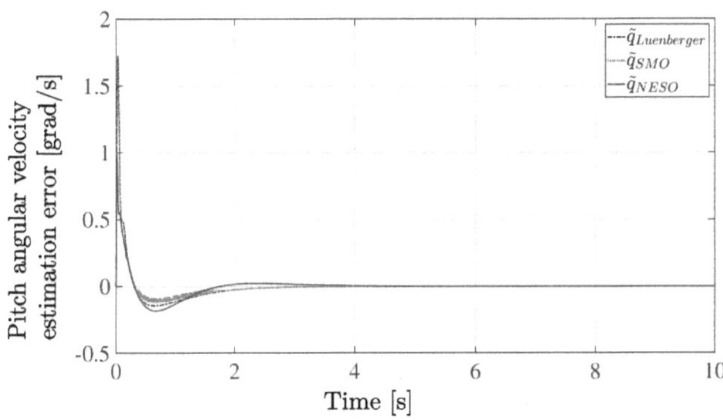

Figure 6.16 Pitch angular velocity estimation error for the three state observers.

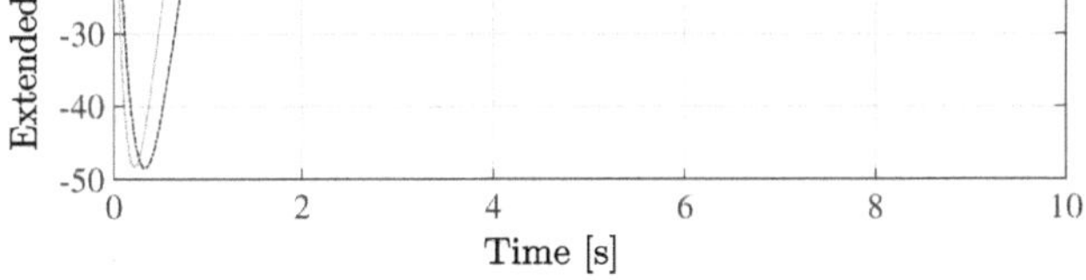

Figure 6.17 Extended state for pitching motion.

For the simulation of the movement of altitude, with the PD control fed back with the states estimated by the observed state, it was proposed that the dynamics start from zero (taking into account that the UAV is at a certain altitude at the beginning of the simulation) and that the desired altitude was 5 meters. The results are shown in Fig. 6.18.

The control signal for the altitude movement with PD control law is shown in Fig. 6.19.

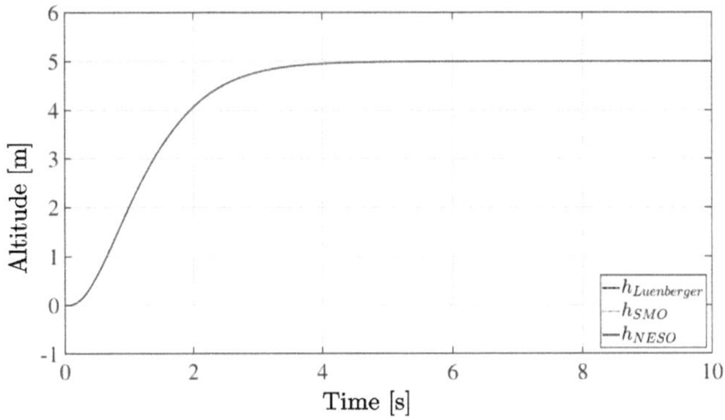

Figure 6.18 Simulation of the altitude movement with PD control for the three observers.

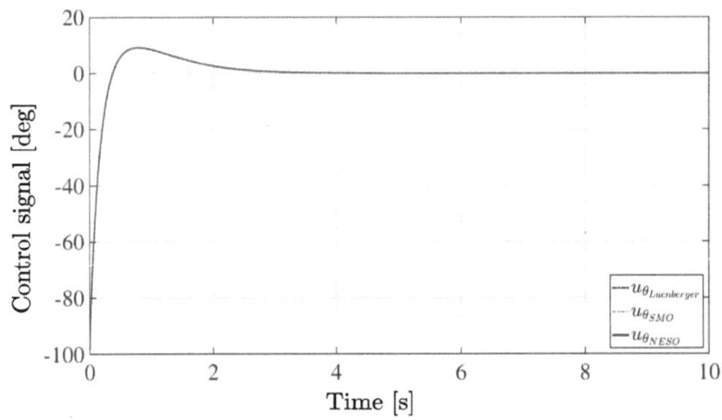

Figure 6.19 PD control signal for pitch motion.

6.6.6 Roll motion simulations with backstepping control law with observers

Figs. 6.20 and 6.21 show the simulation results of three state observers, Luenberger, sliding, and nonlinear extended state modes, that feed back to the backstepping control law for the roll angle and roll angular velocity, respectively.

Figs. 6.22 and 6.23 show the simulation results of the roll angle and roll angular velocity estimation errors, respectively, with backstepping control law.

The extended state of NESO applied to the backstepping control law roll motion is shown in Fig. 6.24.

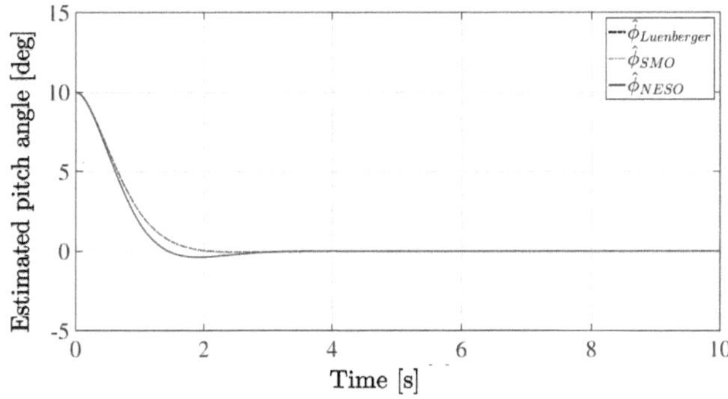

Figure 6.20 Roll angle estimation for the three state observers.

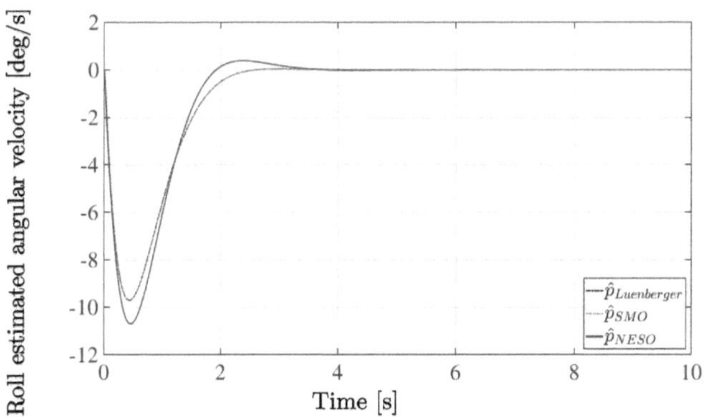

Figure 6.21 Roll angular velocity estimate for the three state observers.

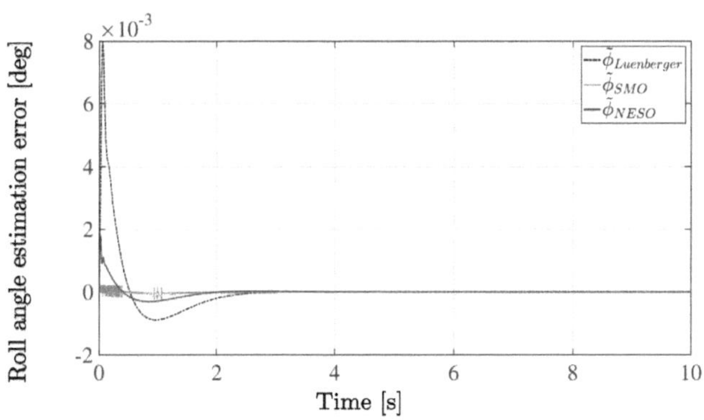

Figure 6.22 Roll angle estimation error for the three state observers.

Fig. 6.25 shows the simulation results of the backstepping control signal applied to the three studied observers.

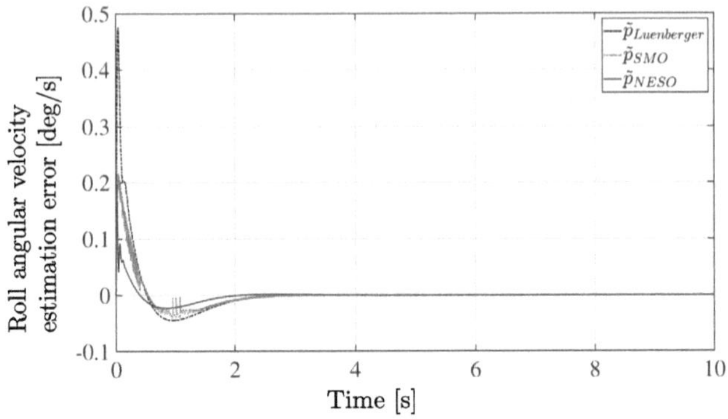

Figure 6.23 Roll angular velocity estimation error for the three state observers.

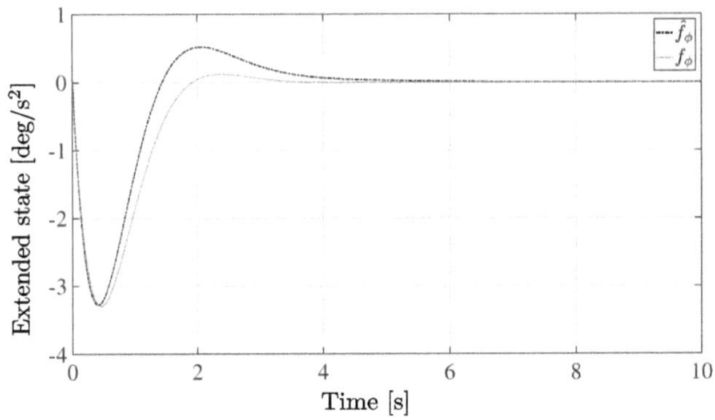

Figure 6.24 Extended state for roll motion.

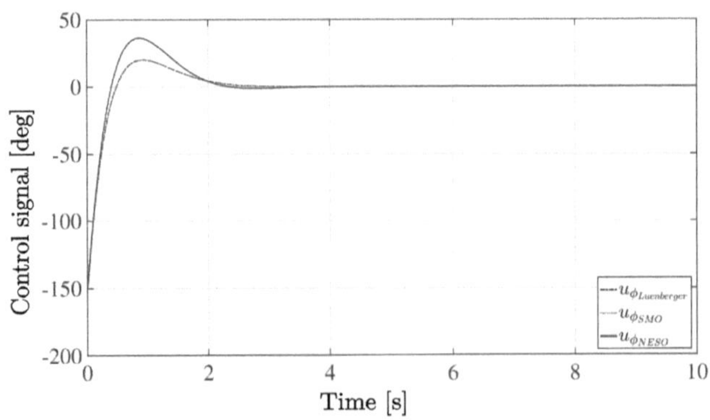

Figure 6.25 Backstepping control signal for roll motion.

6.6.7 Yaw motion simulations with backstepping control law with observers

Figs. 6.26 and 6.27 show the simulation results for the yaw angle and yaw angular velocity, respectively, both for the three observers and the backstepping control law. Figs. 6.28 and 6.29 show the simulation graphs of the estimated errors of the three observers with backstepping control law. Fig. 6.30 shows the estimation of the extended state in simulation with the yaw dynamics and the backstepping control law. The backstepping control signal for the yaw movement is shown in Fig. 6.31.

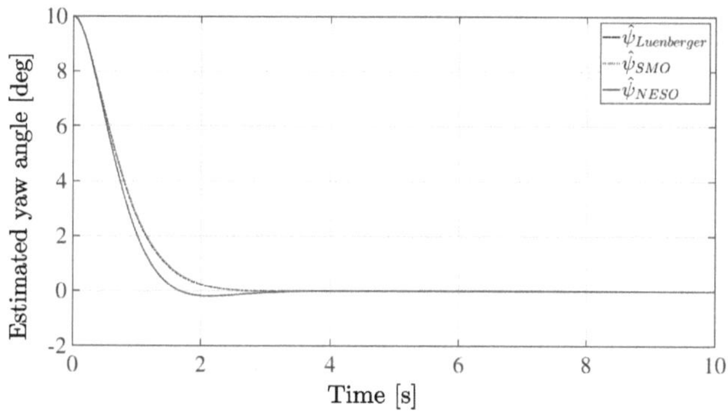

Figure 6.26 Yaw angle estimation for the three state observers.

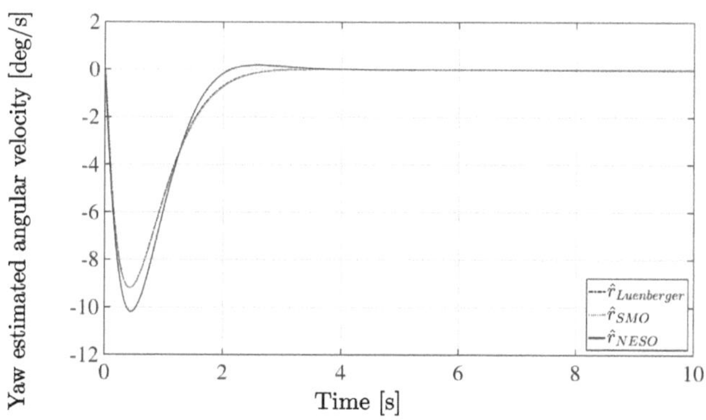

Figure 6.27 Yaw rate estimation for the three state observers.

6.6.8 Altitude movement simulations with backstepping control law with observers

In Fig. 6.32 the altitude simulation results for the three observers and the backstepping control law are shown.

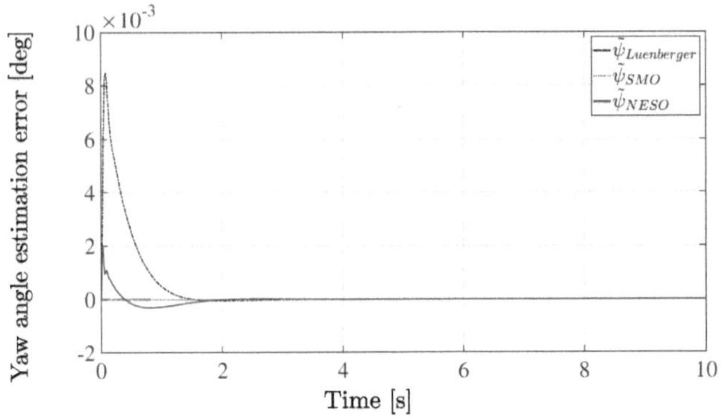

Figure 6.28 Yaw angle estimation error for the three state observers.

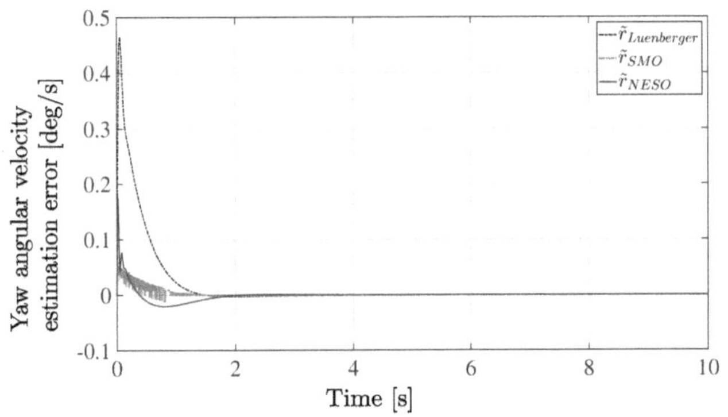

Figure 6.29 Yaw angular rate estimation error for the three state observers.

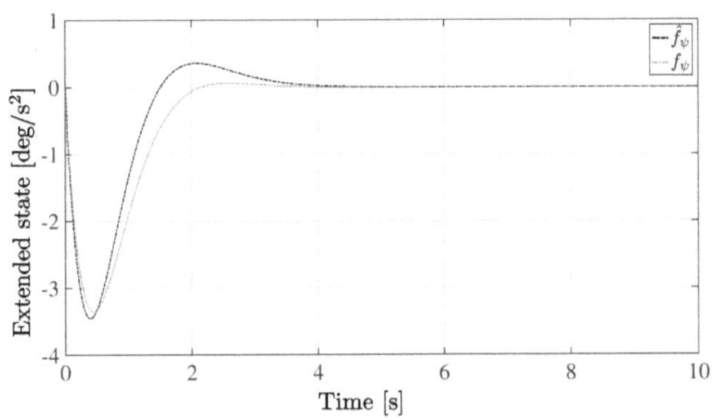

Figure 6.30 Extended state for yaw motion.

Figs. 6.32 and 6.33 show the simulation results for the yaw angle and yaw angular velocity, respectively, both for the three observers and the backstepping control law. In Figs. 6.34 and 6.35 the simulation graphs of the estimated errors of the three ob-

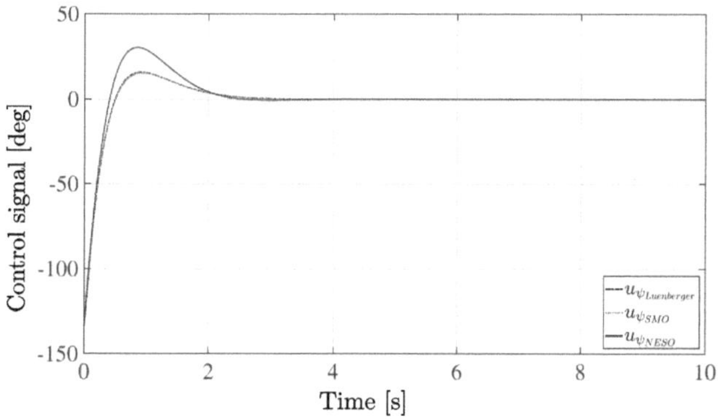

Figure 6.31 Backstepping control signal for yaw movement.

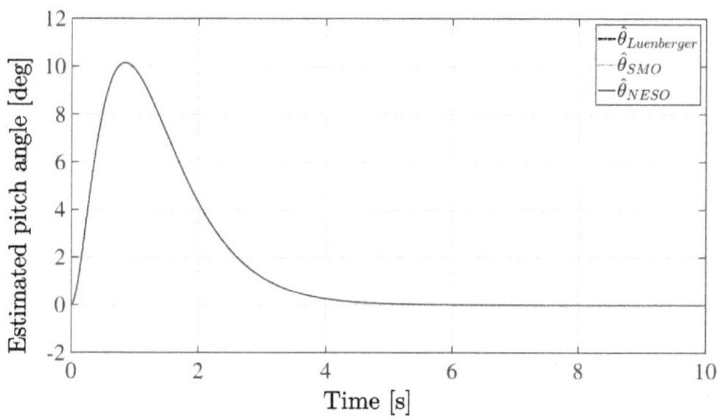

Figure 6.32 Pitch angle estimation for the three state observers.

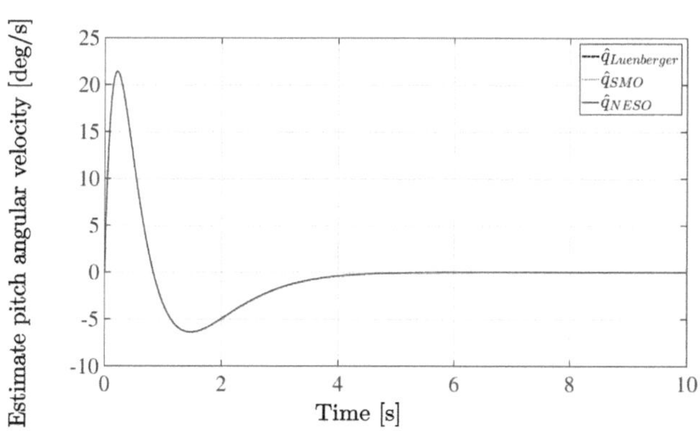

Figure 6.33 Pitch angular velocity estimates for the three state observers.

servers with backstepping control law are shown. Fig. 6.36 shows the estimation of the extended state in simulation with the yaw dynamics and the backstepping control

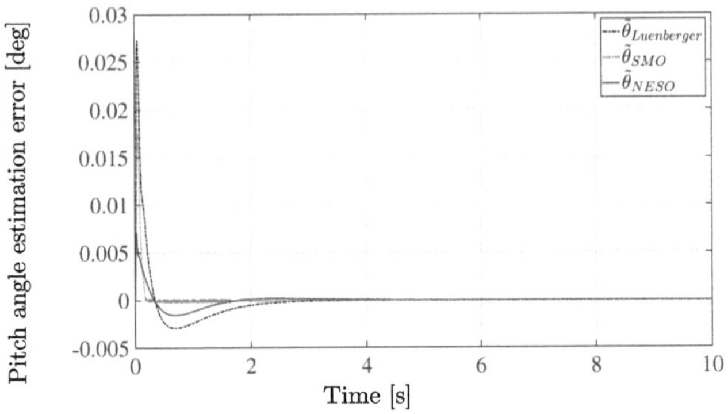

Figure 6.34 Pitch angle estimation error for the three state observers.

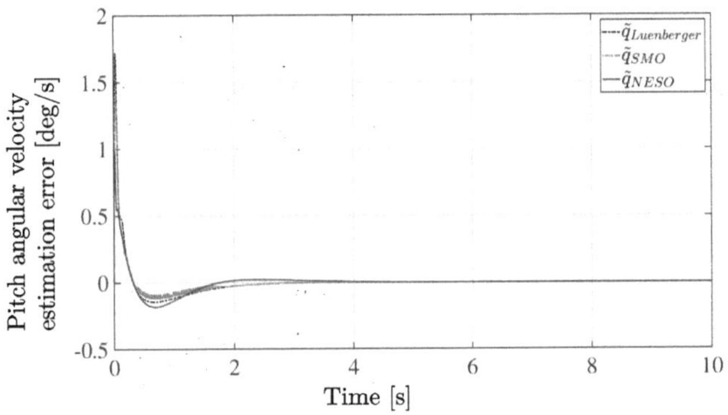

Figure 6.35 Pitch angular velocity estimation error for the three state observers.

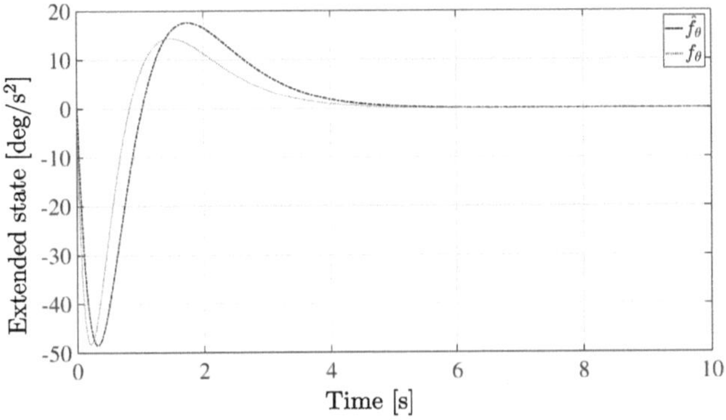

Figure 6.36 Extended state for pitching motion.

law (see Fig. 6.37). The backstepping control signal for the desired pitch is shown in Fig. 6.38. See Table 6.1.

Table 6.1 $\mathcal{L}_2$ norm for the errors in the observers with different control laws.

Controller	**Observer**	**Pitch [deg]**	
		$\mathcal{L}_2[\tilde{\theta}]$	$\mathcal{L}_2[\tilde{q}]$
PD	Luenberger	7.4384e−05	0.0040
PD	SMO	3.9026e−05	0.0033
PD	NESO	2.1874e−05	0.0027
Backstepping	Luenberger	5.8393e−05	0.0031
Backstepping	SMO	1.9963e−05	0.0024
Backstepping	NESO	1.7831e−05	0.0022

Controller	**Observer**	**Yaw [deg]**	
		$\mathcal{L}_2[\tilde{\phi}]$	$\mathcal{L}_2[\tilde{p}]$
PD	Luenberger	2.8827e−05	0.0015
PD	SMO	1.2846e−06	8.9948e−04
PD	NESO	5.5272e−06	4.4866e−04
Backstepping	Luenberger	2.6158e−05	0.0014
Backstepping	SMO	1.0655e−06	8.2088e−04
Backstepping	NESO	5.0468e−06	4.0466e−04

Controller	**Observer**	**Roll [deg]**	
		$\mathcal{L}_2[\tilde{\psi}]$	$\mathcal{L}_2[\tilde{r}]$
PD	Luenberger	4.3017e−05	0.0022
PD	SMO	3.9781e−07	2.5759e−04
PD	NESO	6.6230e−06	4.6950e−04
Backstepping	Luenberger	3.7438e−05	0.0019
Backstepping	SMO	2.8633e−07	2.2965e−04
Backstepping	NESO	5.4340e−06	3.7935e−04

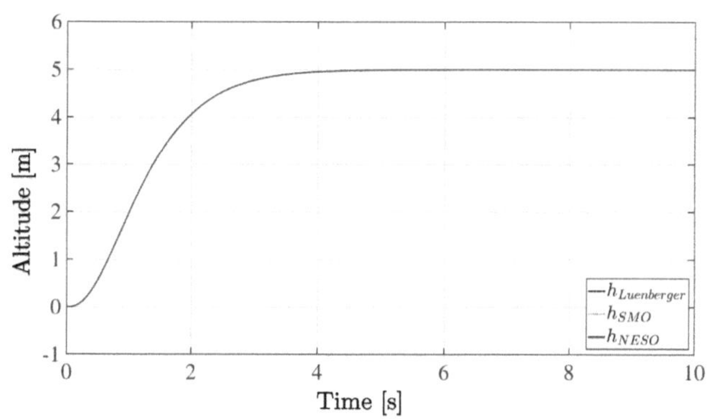

Figure 6.37 Altitude movement simulation with backstepping control for the three observers.

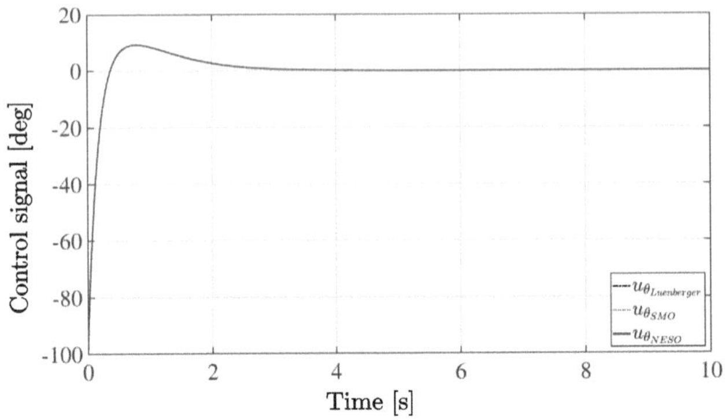

Figure 6.38 Backstepping control signal for pitch motion.

6.7 Concluding remarks

This chapter has described the design of control laws in a closed loop with linear and nonlinear state observers. The development of the control laws with the observers is with the aerodynamic model that defines the fixed-wing UAV shown in Chapter 3. After, the simulation results are shown with the PD and backstepping controllers with the observers Luenberger, SMO, and NESO.

It is worth mentioning that the control laws in the closed loop with the observers are realized for the three fundamental movements for the flight of a fixed-wing UAV, altitude, yaw angle, and roll angle.

Testbed and experimental results 7

Fixed-wing UAVs are fascinating vehicles to implement linear and nonlinear control laws. Nowadays, a great variety of this type of aircraft are offered to perform automatic flights. It is inconvenient to program different control laws in these systems because the manufacturers or developers do not allow program change this type of systems, which is known as a closed structure or closed architecture.

Others in the UAV market can be programming and testing other control techniques. Still, due to their software and hardware structure, some of these systems do not support some robust control laws, nonlinear control techniques, and state observers. Based on the technical disadvantages presented before, we have developed a fixed-wing UAV that is able to support hardware and software. The controllers and observers are implemented with robust algorithms.

This chapter is divided into the following topics. Section 7.1 describes the experimental platform; in this testbed some control laws and state observers mentioned in the previous chapters are programmed. Section 7.2 presents the brushless and servo motors used in fixed-wing UAVs. The inertial measurement used in the embedded system is presented in Section 7.3. Sections 7.4–7.8 show the electronics systems that complement the autopilot system. Finally, Section 7.10 presents the results obtained in real flight tests with the linear and nonlinear controllers and state observers.

7.1 Experimental testbed

For the development of the experimental testbed (fixed-wing UAV), an aeromodelism model is considered, which corresponds to a model airplane *T-28 Trojan*. The parameters of the airplane are shown in Table 7.1.

Table 7.1 Parameters of the airplane of aeromodelism *T-28 Trojan*.

Parameter	Value
Wingspan	1.118 m
Fuselage	0.914 m
Wing area	0.09 m^2
Approximate weight	1500 kg
Motors	Aileron, elevator, rudder, front engine

Design of Control Laws and State Observers for Fixed-Wing UAVs. https://doi.org/10.1016/B978-0-32-395405-1.00016-9

7.2 Motors or actuators in a testbed

The testbed uses four motors; one is for the throttle located in the front of the nose of the fixed-wing UAV (see Fig. 7.1). For the brushless motor a 40-Amp electronic stability control unit is necessary. The other three are servo motors to move the control surfaces: elevator, rudder, and aileron (see Fig. 7.2).

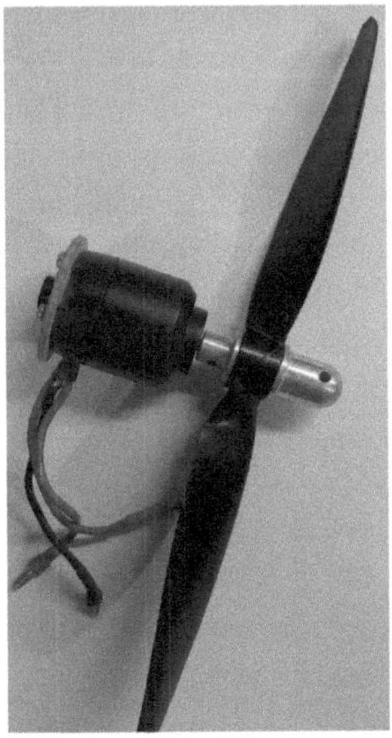

Figure 7.1 Brushless motor.

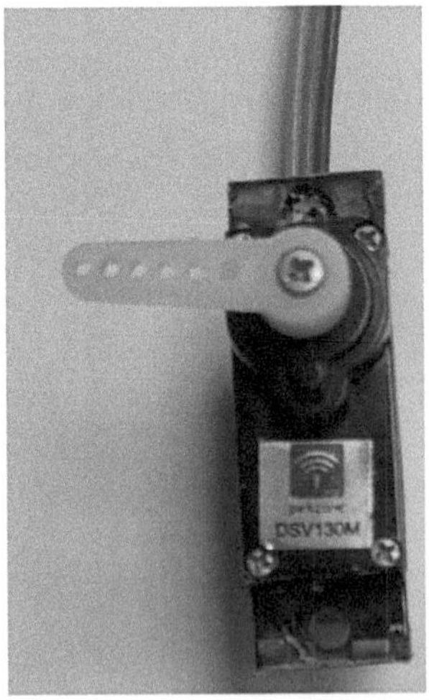

Figure 7.2 Example of servo motor used for elevator, rudder, and aileron.

7.3 Inertial measurement unit (IMU)

The fixed-wing UAV has an IMU on board, which provides the Euler angles and the angular rates of the airplane. The used IMU is the *Microbotics MIDG II*. This IMU is power supplied with 9 V of direct current; see Fig. 7.3. The Euler angles and angular rates from the IMU are obtained with the Recommended Standard 232 (RS232) protocol. The IMU needs 9 V of direct current to have a perfect performance; for this a Traco Power DC-DC converter is used.

Figure 7.3 Inertial measurement unit MIDG II.

7.4 Telemetry

For the telemetry, the Xbee modules of the series *XBee-PRO S2B* are used. These modules have a range of 1.5 km and are power supplied with 3.3 V of direct current; see Fig. 7.4.

Figure 7.4 Xbee to obtain the variables from the fixed-wing UAV.

7.5 Optocoupler

We use an optoisolator, also called an optocoupler, which is a component that transfers electrical signals between two isolated circuits by using light. The used optocoupler prevents high voltages from affecting the system receiving the signal. With this electronic component, we have isolated the electronic control card of the servo motors to avoid damage.

In Fig. 7.5 the circuit configuration which is used in the electronic controller card is presented. The values of the resistors R_1 and R_2 are 100 Ω and 1 kΩ, respectively. The input signal is sent from the microprocessor *Rabbit 6000*, and the output signal is sent to the servo motor input. The signal received by the servo motor has an amplitude of 5 V.

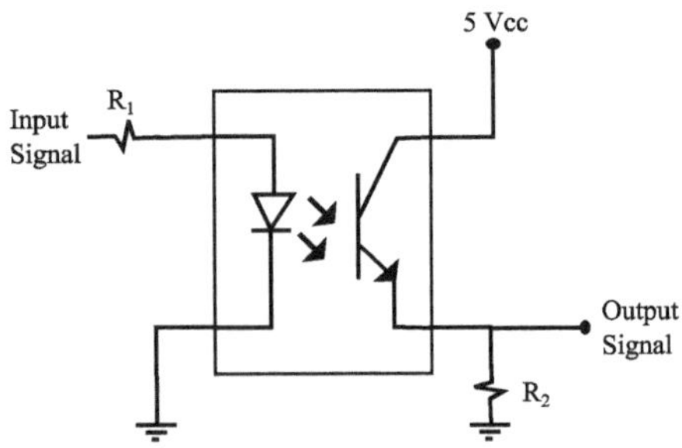

Figure 7.5 Optocoupler.

7.6 Microcontroller and altimeter

The fixed-wing UAV has an altimeter *MS5607* mounted on one of its wings, which is connected to a microcontroller *Propeller* (see Fig. 7.6). Both are development by *Parallax Inc*. The microcontroller is used exclusively to read the data from the altimeter, and subsequently, the microcontroller sends the data to the microprocessor *Rabbit*.

The communication protocol between the altimeter and the microcontroller is I^2C, and the communication protocol between the microcontroller and the microprocessor is *RS232*. The altitude sensor and the microcontroller are power supplied with 5 V of direct current.

The P8X32A QuickStart is a simple and accessible development platform. It is fully expandable and provides unimpeded access to all I/O pins through an expansion header but includes some button inputs and LEDs to demonstrate programming. Let us present some key features of this microcontroller:

- Expansion header provides access to Vss, Vdd, 32 I/O pins.
- The 5-MHz crystal on board may be disabled by removing the SMT jumper, allowing replacement with an external through-hole crystal.
- Eight resistive touch buttons.
- Eight buffered LEDs.

- It can be used with the altimeter *MS5607*.

The fixed-wing UAV has an altimeter *MS5607* mounted on one of its wings (see Fig. 7.7), which has a high resolution of 20 cm. It gives highly accurate altitude, pressure, and temperature readings. This altitude module includes a high-linearity barometric pressure sensor and a high-resolution temperature output, allowing an altimeter/thermometer implementation without any additional sensors. Different operation modes enable the user to optimize conversion speed or current consumption. Some key features of this module are:

- 24-bit pressure and temperature values.
- A wide pressure range.
- Successfully tested at 120000 feet.

Figure 7.6 Microcontroller (Propeller P8X32A).

Figure 7.7 Altimeter (MS5607).

7.7 Microprocessor Rabbit 6000

The microprocessor *Rabbit* 6000 is the central unit of processing to develop the electronic board to program the control laws and the state observers; see Fig. 7.8. Besides, it has the task of sending the control signals to the servo motors, which move the control surfaces of the fixed-wing UAV.

Figure 7.8 Microprocessor Rabbit (RCM6000).

To communicate via *RS232*, the control signals are sent to the servo motors, and the signals are read from the Futaba receiver. A card is built that complements the microprocessor development card *Rabbit* 6000; see Fig. 7.9.

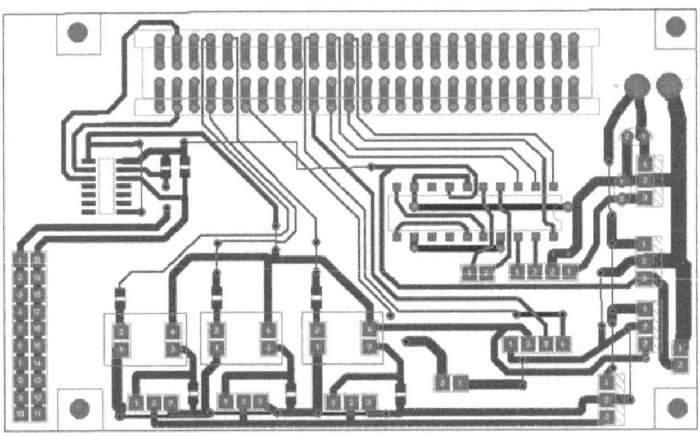

Figure 7.9 PCB complementary for the microprocessor Rabbit (RCM6000).

To fly the airplane the radio exerts control using the *Futaba T7C* controller (transmitter and receiver). With this radio (*Futaba T7C*) the signals are sent to the servo motors in open loop and lead the fixed-wing UAV to take off and landing; see Fig. 7.10. With this same radio control the signal of automatic mode of the fixed-wing UAV is sent, that is, it activates the control action and at that time the autopilot steers the fixed-wing UAV. The receiver of the Futaba radio is powered with 5 V of direct current, while the radio transmitter is powered by a battery of 9.6 V of direct current.

Before sending the signal obtained by the controller towards the servo motors, a linear transformation is applied so that the control signal is within the values allowed by the servo motors and the control surfaces of the fixed-wing UAV are properly deflect.

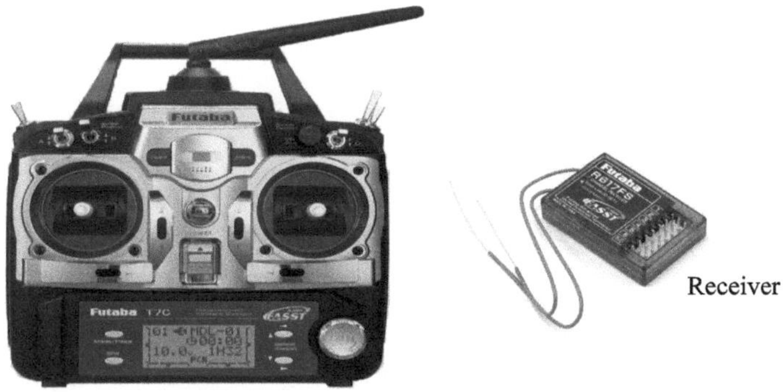

Figure 7.10 Radio control Futaba (T7C).

The linear transformation is given by

$$y = mx + b, \tag{7.1}$$

where m represents the slope, x is the variable, and the value of b determines the point where intersects the y-axis. Fig. 7.11 shows the used straight line to do the transformation in the case of the prototype fixed-wing UAV. The value of ± 200 indicates the duty cycle of the pulse width modulation (PWM) of the microprocessor, which allows the control surfaces to move at their maximum and minimum values, which are ± 20 degrees (deviation of the control surface). Thus, the value obtained in the variable y of Eq. (7.1) is the signal sent to the servo motors of the control surfaces.

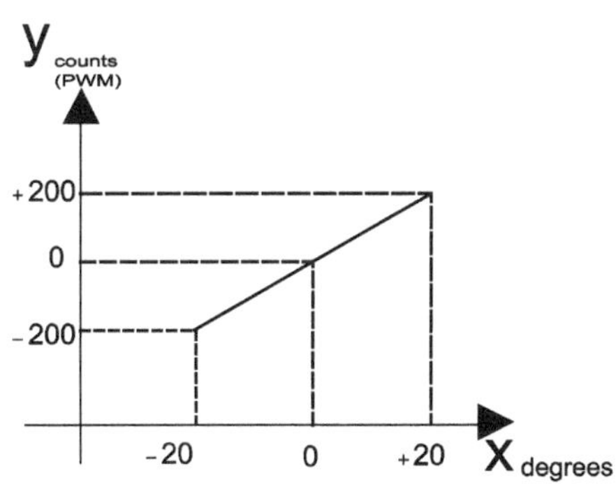

Figure 7.11 Linear transformation.

Fig. 7.12 shows the electronic system mounted on the fixed-wing UAV.

Fig. 7.13 represents the electronic system block diagram used to carry out the real flight test. Then, the embedded system block defines the Rabbit microprocessor where the control laws are programmed. The state observer uses a programming language

Figure 7.12 Embedded system mounted on the fixed-wing UAV.

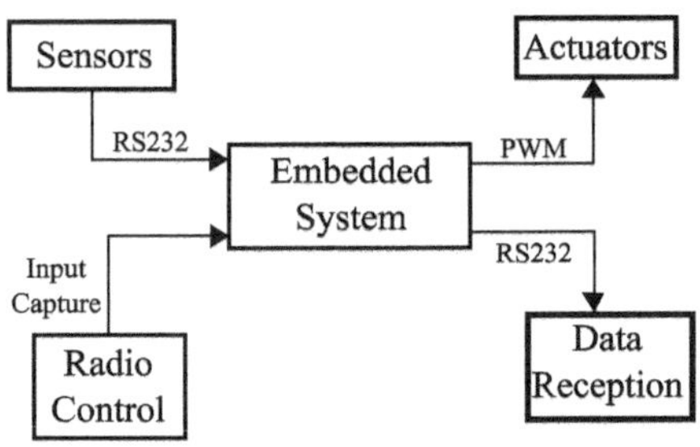

Figure 7.13 Electronic system block diagram.

called Dynamic C, which is mainly used to program the family of Rabbit microprocessors.

The block called actuator represents the servo motors and brushless motor that are moved via PWM. The Rabbit microprocessor sends such PWM pulse. There are three servo motors that drive the control surfaces of the fixed-wing UAV (elevator, rudder, and aileron), and the brushless motor is the airplane thrust.

Data are sent to/received from the fixed-wing UAV, and after graphical analysis, the desired value is necessary for a block called data reception; this block contains the XBee module mounted in the fixed-wing UAV and the station on Earth composed of a computer with an XBee module. For these modules the communication protocol is RS232.

On the other side, the sensors used are an IMU and an altimeter. The second sensor is read first by the Parallax microcontroller, and after this the microcontroller is programmed to send information with an RS232 protocol to the Rabbit microprocessor.

Finally, it is necessary to read the signals from a radio control for take-off and landing (even when the fixed-wing UAV achieves a high altitude). The control law and observers are tested sending or activating a switch in the radio control to start the real flight test with these algorithms.

7.8 Li-po battery

The power source of all electronic embedded systems in this book is a Li-po battery; this battery offers better performance for UAVs. Using a battery with the same characteristics in every flight test is necessary because the flight performance is different if the same discharge battery current is not kept; this is very important to test control laws and observers in a real flight test.

The technical specifics of the battery are 2200 mAh, 11.1 V, and 45–90 discharge cycles (see Fig. 7.14).

Figure 7.14 Li-po battery.

7.9 Experimental results for linear and nonlinear controllers

This section shows the experimental results obtained with the PD controller, the PID controller, the LQR controller, the LQR controller with discrete-time Kalman filter, and nonlinear backstepping control.

7.9.1 PD controller

In Table 7.2 the used gains in the experiments for the PD controller are presented. These gains are tuned by trial and error in the flying field, starting with the values obtained in the simulation results shown in the previous chapters.

In Fig. 7.15 the experimental results obtained with the PD controller applied for altitude are observed. In the upper graph it is observed that the controller is applied

Table 7.2 PD controller gains in experiments.

Movement	k_{pl}	k_{vl} [s]
Altitude	10	0.1
Yaw	5	1
Roll	5	1

at three different altitudes. The acronym AC is used to indicate when the control law has been applied, and the acronym MC indicates when the manual control has been applied, that is, when the fixed-wing UAV was flown by radio control.

It is appreciated in the upper graph of Fig. 7.15 that the altitude of the fixed-wing UAV is maintained very near the reference. In the lower graph of the same Fig. 7.15, the control action applied to accomplish the control objective is shown. The control signal is bounded to ± 20 degrees because these are the allowed values by the control surfaces of the fixed-wing UAV (elevator, rudder, and ailerons).

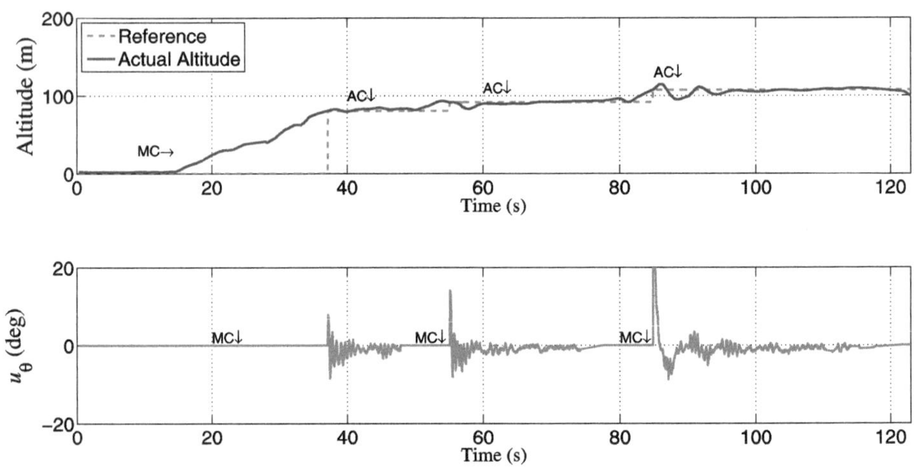

Figure 7.15 Action of the PD controller for altitude.

In addition to the above acronyms, in the yaw movement results, CS and WG mean a change of sign and wing gusts, respectively. In Fig. 7.16 the experimental results obtained with the PD controller applied for the yaw angle are shown. In the upper graph, within 40 seconds, a change of sign is presented; therefore, the IMU for the yaw angle has an operation range of ± 180 degrees.

For the yaw angle, we seek to apply control laws in the range of ± 100 degrees to avoid the change of sign that the IMU produces. In the upper graph of Fig. 7.16, the control law PD is applied for yaw at different angles of reference. It is appreciated that the control objective is achieved. The second time that the controller (CA) was applied, it was appreciated that the fixed-wing UAV was perturbed by wind gusts. However, the yaw angle stays close to the desired reference.

Besides, the third time that it was applied, the controller perceived wind gusts more frequently; however, the yaw angle was maintained or stayed close to the reference. The lower graph of Fig. 7.16 presents the control action which the PD controller applies for the yaw angle; it is observed that when the wind gusts are present in the flight test, the control law applied a higher control action to lead the fixed-wing UAV again to the reference desired (see the lower graph of Fig. 7.16).

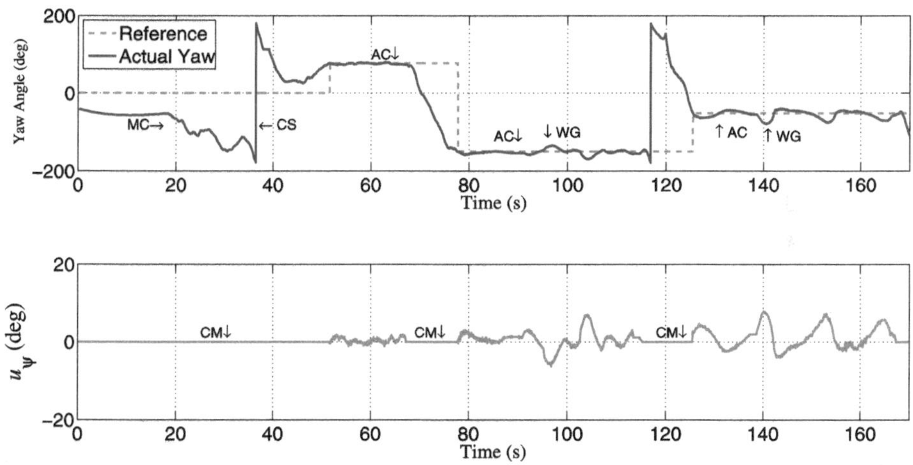

Figure 7.16 Action of the PD controller for yaw.

For the roll movement, the control objective consisted in keeping the roll angle at 0 degrees. The experimental results obtained with the PD controller in the roll are shown in Fig. 7.17. The upper graph of this figure shows how the roll angle stays close to the reference by applying the PD controller. The second time the controller was applied, there were perturbations by wind gusts, and in the lower graph of Fig. 7.17 it is appreciated that a higher control action is applied to lead the fixed-wing UAV to the reference value.

The third time the controller was applied, there were no disturbances and it is appreciated in the lower graph of Fig. 7.17 that a lower control action was applied when there were no wind gusts.

7.9.2 PID controller

Table 7.3 shows the used gains in the experimental tests with the PID controller. The gains were obtained in a heuristic form in the flying field, considering the gains obtained in simulations at the beginning of the test.

The results obtained with the PID controller for altitude are shown in Fig. 7.18. In the upper graph of the same figure the altitude references are shown. It is observed that the actual altitude of the fixed-wing UAV oscillates about the reference. A higher control action is applied when oscillations are present (see the lower graph of Fig. 7.18).

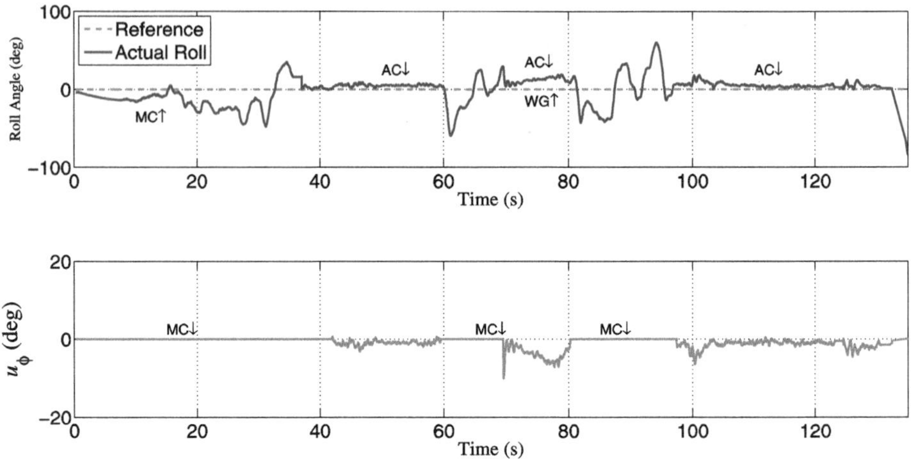

Figure 7.17 PD controller for roll.

Table 7.3 PID controller gains in experiments.

Movement	k_{pl}	k_{il} [1/s]	k_{vl} [s]
Altitude	10	0.00001	0.1
Yaw	5	0.01	1
Roll	5	0.01	1

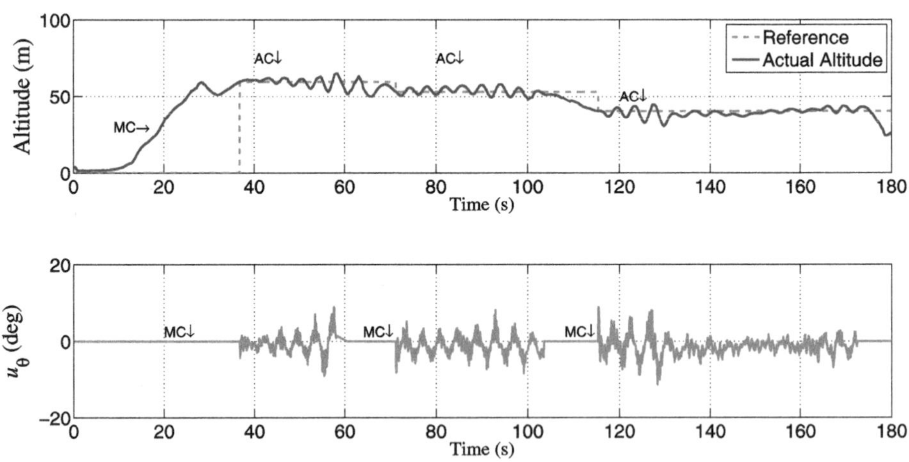

Figure 7.18 Action of the PID controller for altitude.

It is observed that the better response with the PID controller was obtained the third time this controller was applied, between 140 and 160 seconds.

Fig. 7.19 shows the results obtained with the PID controller for yaw. In the upper graph it is observed how the controller tried to keep the yaw angle close to the desired reference. The first time that it was applied, it was perturbed by wind gusts. In the presence of these gusts, the controller led the fixed-wing UAV back to the desired reference value. It is also observed that the fourth time the controller was applied, the yaw angle stayed on the reference for a long time. The second graph of Fig. 7.19 presents the control action to achieve the control objective.

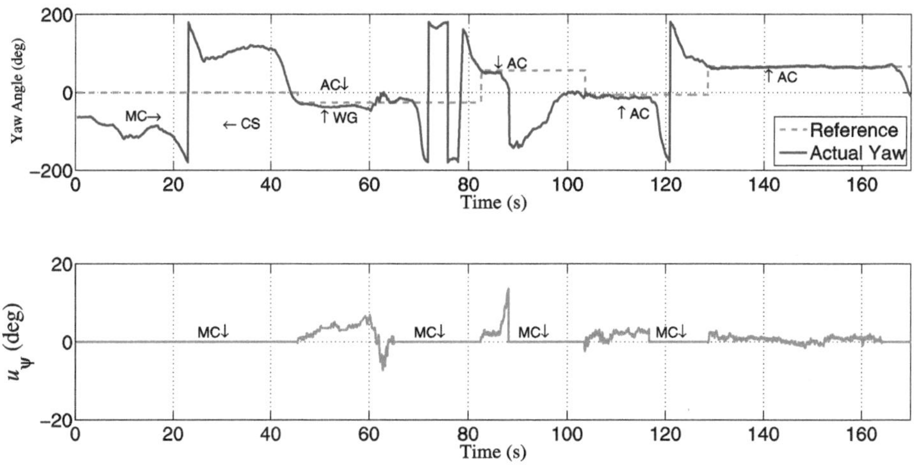

Figure 7.19 Action of the PID controller for yaw.

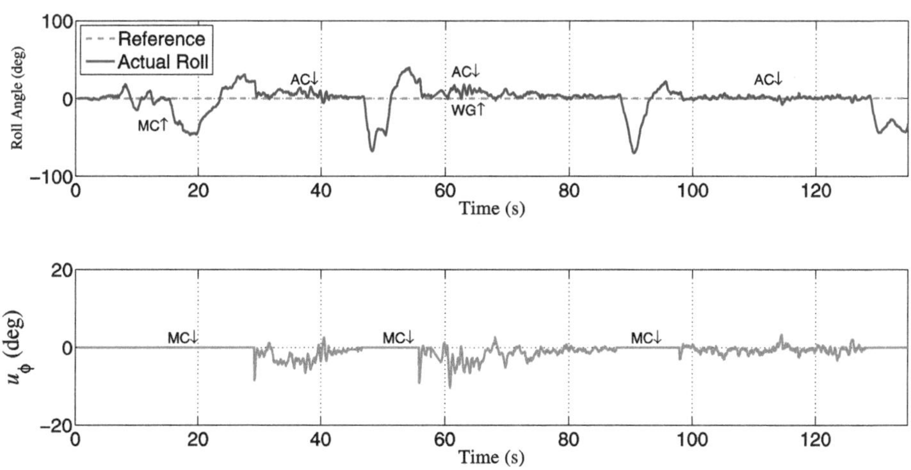

Figure 7.20 Action of the PID controller for roll.

The results obtained with the PID controller for roll are shown in Fig. 7.20. The lower graph of Fig. 7.20 presents the control action applied by the PID controller for roll. In the upper graph, it is appreciated how the PID controller led the roll angle close to the reference. The second time the control law was applied, there were perturbations by wind gusts. It is observed that the controller tries to keep the roll angle near the reference of 0 degrees. Besides, the PID controller obtained the best control response the third time that the automatic control was applied.

7.9.3 LQR controller

The used gains in the experiments of the LQR are the same as used in the simulations; these gains are shown in Table 4.4.

Fig. 7.21 shows the results obtained in experiments with the LQR controller. In the upper graph, it is observed that the altitude of the fixed-wing UAV is maintained or stays near the references assigned. The lower graph of Fig. 7.21 presents the control action applied by the LQR. It is worth mentioning that the results shown in Fig. 7.21 were obtained with the optimal gains shown in Table 4.4. It is possible to obtain a better response of the controller by retuning the calculated gains, but here the important message is the stabilization of the fixed-wing UAV, or at least it is maintained near the desired references.

The results obtained with the LQR controller applied for the yaw angle are shown in Fig. 7.22. In the upper graph, it is shown that the LQR controller led the yaw angle to the reference, but the first time the automatic control was applied, the yaw angle was not near the reference.

The second time the control law was applied, with wind gusts, the same response was observed. The calculated gains are low for these perturbations, such that the correction of the error is slow.

The third time it was applied, the controller perceived much weaker wind gusts, which allows better performance of the LQR controller (see the upper graph of Fig. 7.22). In the lower graph of Fig. 7.22, the control actions are presented that apply whenever the automatic control is used. It is observed that when the error is big, also

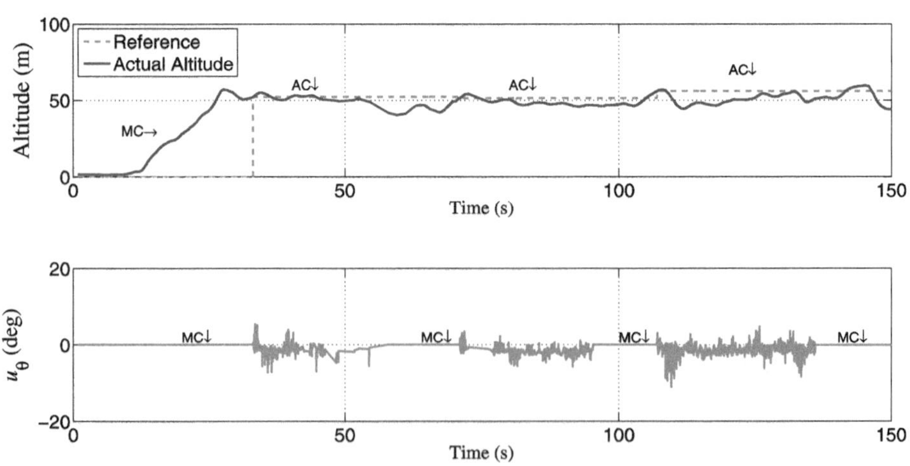

Figure 7.21 Action of the LQR controller for altitude.

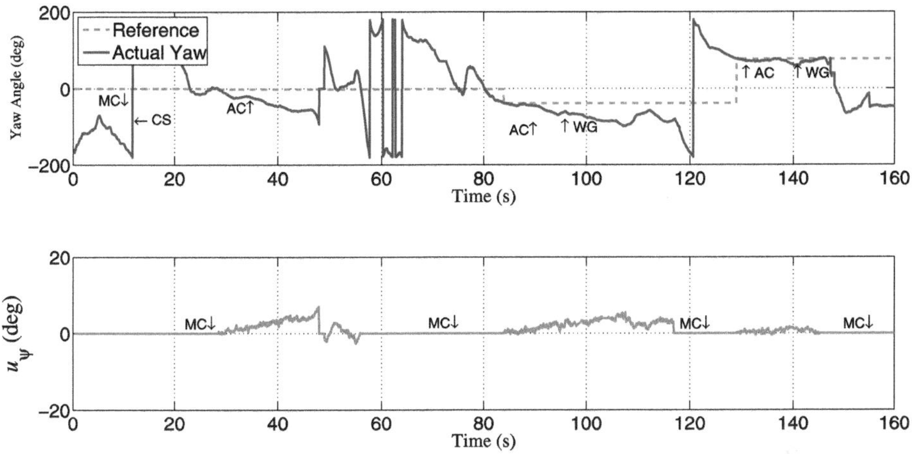

Figure 7.22 Action of the LQR controller for yaw.

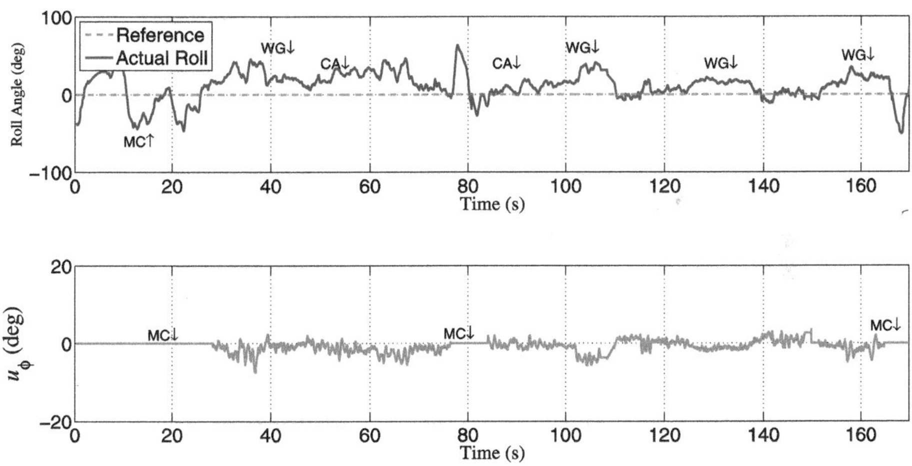

Figure 7.23 Action of the LQR controller for roll.

the control action increases, but it needs more control action to correct in less time, and the system does not become unstable.

The results obtained with the LQR controller for the roll angle are presented in Fig. 7.23. It is appreciated in the upper graph that when the fixed-wing UAV is perturbed by wind gusts, the error of the roll angle is big, but it is worth mentioning that these experiments are performed with the gains calculated with the methodology of the LQR controller; therefore, it is possible to improve these results by retuning the controller gains. The objective of these tests maintains the calculated gains, proving that the LQR controller at least tries to keep the roll angle on or near the reference.

It is proven that the LQR controller for roll in the presence of perturbations has no good response time; that is, the LQR controller for roll takes a lot of time to bring back the roll angle near or above the desired reference of 0 degrees. In the lower graph of Fig. 7.23 the control action which the LQR controller applies for the roll angle is presented.

7.9.4 LQR controller with discrete-time Kalman filter

The gains used in the experiment tests for the LQR controller with discrete-time Kalman filter are the same as used in the simulations; see Table 4.4 and Table 4.3, respectively.

The obtained results for altitude by applying the LQR controller with discrete-time Kalman filter are presented in Fig. 7.24. In the upper graph, it is observed how the altitude of the fixed-wing UAV is maintained around the desired references. Some oscillations are presented in the fixed-wing UAV in trying to stay close to the reference. In the lower graph of Fig. 7.24 the control actions applied by the LQR controller with discrete-time Kalman filter are shown.

The experiment applying the LQR controller with discrete-time Kalman filter for the yaw angle was performed taking as reference 0 degrees. It is appreciated that the controller has no good performance with the previously calculated gains with the LQR technique.

It is appreciated that the fixed-wing UAV tends to oscillate with wind gusts. These oscillations did not have a considerable magnitude to render the system unstable; besides, oscillations took place near or above the reference; see the upper graph of Fig. 7.25. This response can be improved by fine-tuning the gains, but in this experiment, the important message is to do the flight tests with the calculated gains with the LQR controller. In the lower graph of Fig. 7.25, the control action applied by the LQR with discrete-time Kalman filter is presented. The magnitudes of the oscillations of the control action are appreciated.

Fig. 7.26 shows the experimental results obtained applying the LQR controller with discrete-time Kalman filter for the roll angle. The reference to which the roll angle must converge is 0 degrees. In the upper graph of Fig. 7.26, it is observed that the roll angle presents oscillations, but tries to stay close to the reference when the controller is applied. Besides, the first time it was applied, the controller perceived some perturbations by wind gusts. The lower graph of Fig. 7.26 shows the control action that is applied to achieve the control objective.

7.9.5 Backstepping controller

In Table 7.4 the used gains in the experiments for the backstepping controller are shown. The tuning of the gains is realized in a heuristic form, that is, by trial and error.

The backstepping controller presents an excellent performance for altitude, and this controller keeps the altitude of the fixed-wing UAV above the reference most of the

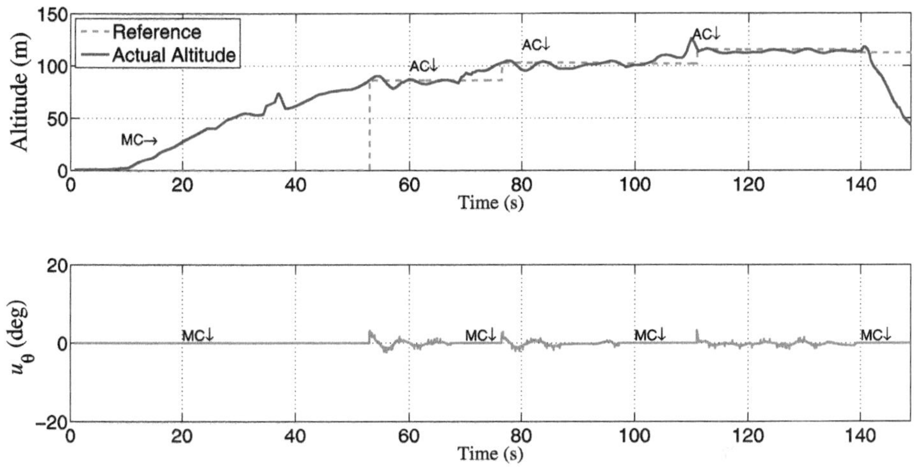

Figure 7.24 Response of the LQR controller with discrete-time Kalman filter for altitude.

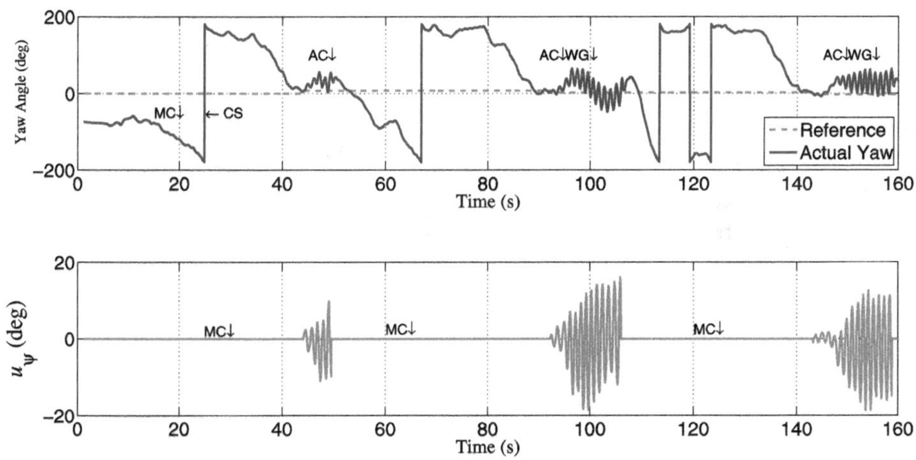

Figure 7.25 Response of the LQR controller with discrete-time Kalman filter for yaw.

Table 7.4 Gains of the backstepping controller in experiments.

Movement	k_{1l} [1/s]	k_{2l} [1/s]	k_{3l} [1/s^2]
Altitude	0.1	0.5	4
Yaw	0.1	0.5	2
Roll	0.1	1	3

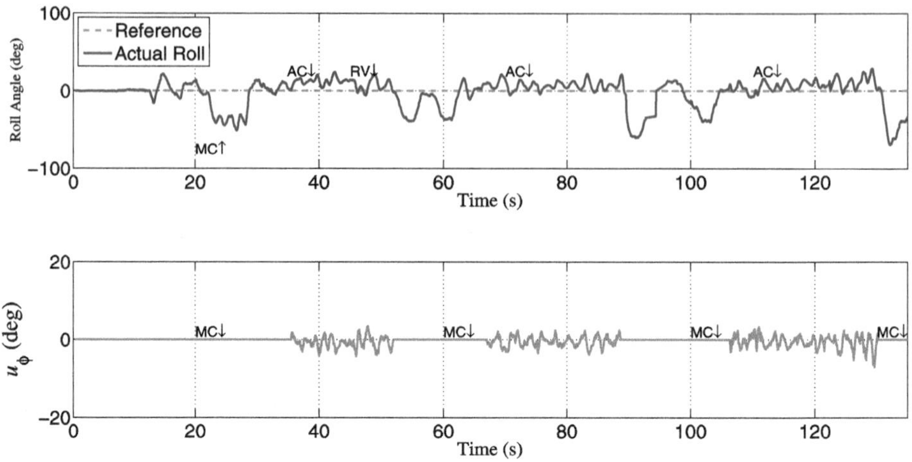

Figure 7.26 Response of the LQR controller with discrete-time Kalman filter for roll.

time; see the upper graph of Fig. 7.27. Besides, in comparison with the other controllers that are shown in this chapter (real-time implementation), the backstepping controller applied a small control action to maintain the fixed-wing UAV on the desired reference; see the lower graph of Fig. 7.27.

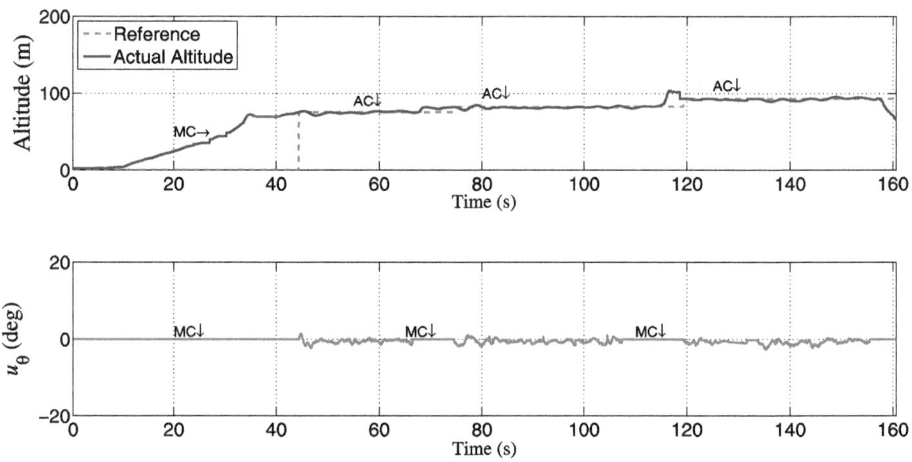

Figure 7.27 Response of the backstepping controller for altitude.

The obtained results for the yaw angle applying the backstepping controller are shown in Fig. 7.28. In the upper graph of this figure, it is observed that the control law has good performance because the yaw angle is maintained close to the desired reference. The lower graph of Fig. 7.28 shows the applied control action, which was lower in comparison with the other controllers presented in this chapter for the altitude control.

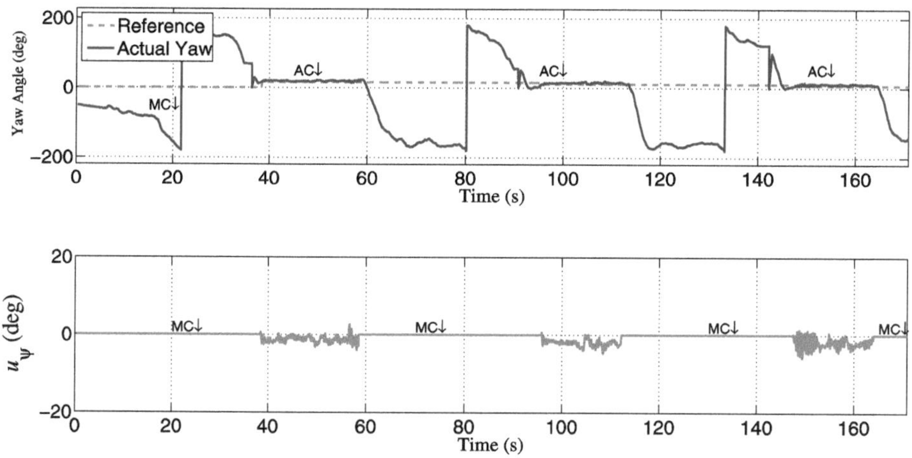

Figure 7.28 Action of the backstepping controller on yaw.

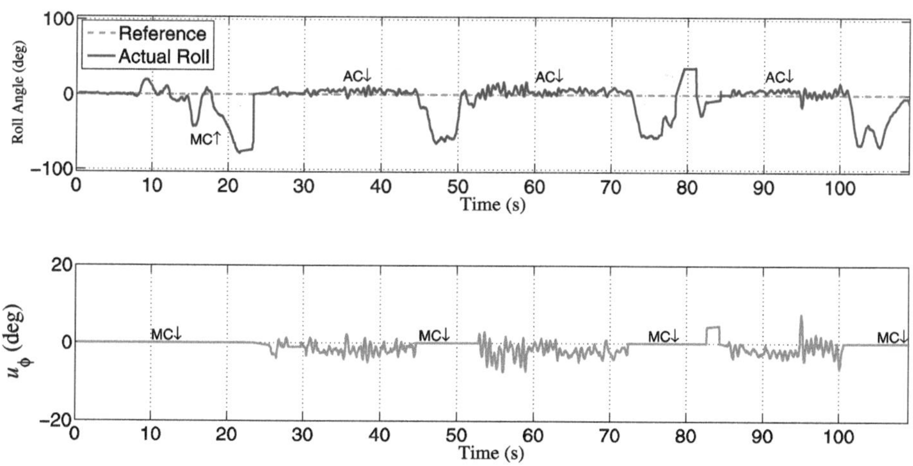

Figure 7.29 Action of the backstepping controller on roll.

Fig. 7.29 shows the obtained results with the backstepping controller for roll; in the upper graph it is appreciated that the roll angle is maintained close to the reference, and the third time the controller was applied, in a time interval the roll angle remained around the desired reference. In the lower graph of Fig. 7.29, the control action for roll is shown. It is appreciated that the control signal is not as small as with the other controllers. See Table 7.5.

Table 7.5 $\mathcal{L}_2$ norm for the errors of the linear and nonlinear controllers in the real flight test.

Controller	Altitude [m]		Yaw [deg]		Roll [deg]	
	$\mathcal{L}_2[e_\theta]$	$\mathcal{L}_2[u_\theta]$	$\mathcal{L}_2[e_\psi]$	$\mathcal{L}_2[u_\psi]$	$\mathcal{L}_2[e_\phi]$	$\mathcal{L}_2[u_\phi]$
PD	14.9513	0.4489	13.4787	0.2648	10.3812	0.2194
PID	8.8783	0.3841	11.8016	0.2130	2.4997	0.2322
LQR	8.3176	0.2436	11.6437	0.2332	2.9985	0.2154
LQR-DKF	15.6543	0.0777	14.8595	0.4923	2.8884	0.1568
Backstepping	13.9771	0.1262	14.6339	0.1615	3.1152	0.2367

7.10 Experimental results for linear and nonlinear observers

This section presents the experimental results obtained with the linear and nonlinear observers, including three state observers applied to fixed-wing UAV control, which implements two control laws (PD, backstepping) in yaw, pitch, and roll motions. For the practical implementation of state observers studied, an experimental platform that is reported in Section 7.1 was used.

7.10.1 Luenberger observer applied to a fixed-wing UAV with PD control law

The experimental results of the Luenberger observer for pitch, roll, yaw, and altitude movement of the UAV are presented below.

Pitch motion

Figs. 7.30 and 7.31 show the experimental results of the Luenberger observer and a PD control law for pitch angle and pitch angular velocity, respectively. Figs. 7.32 and 7.33 show the angle estimation and angular velocity errors. Fig. 7.34 shows experimental results of altitude control. Fig. 7.35 shows the control signal of the same experiment.

Roll motion

Figs. 7.36 and 7.37 show the experimental results of the Luenberger observer and a PD control law for the roll angle and roll angular velocity, respectively. Figs. 7.38 and 7.39 show the estimation angle and angular velocity errors. Fig. 7.40 shows the control signal from the same experiment.

Yaw motion

Figs. 7.41 and 7.42 show the experimental results of the Luenberger observer and a PD control law for the yaw angle and yaw angular velocity, respectively. Figs. 7.43 and 7.44 show the angle estimation and angular velocity errors. Fig. 7.45 shows the control signal of the same experiment.

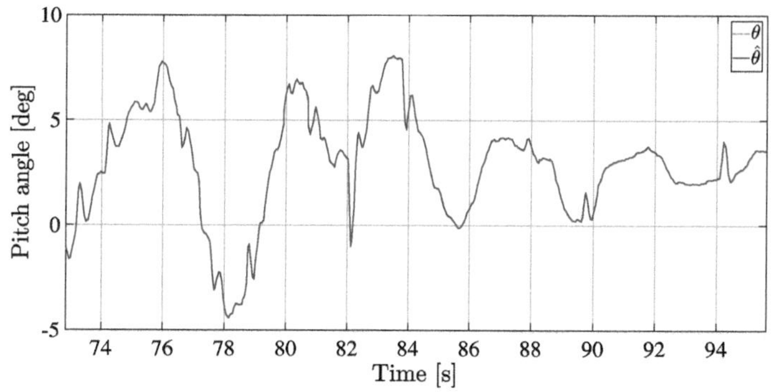

Figure 7.30 Results of the experiment applying the Luenberger observer to a fixed-wing UAV for pitch angle.

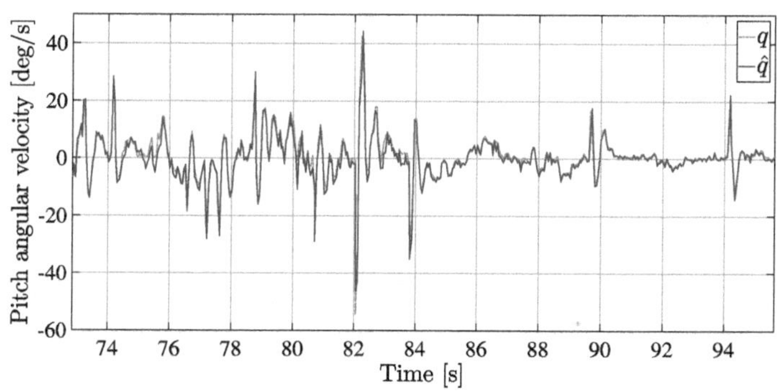

Figure 7.31 Results of the experiment applying the Luenberger observer to a fixed-wing UAV for pitch angular velocity.

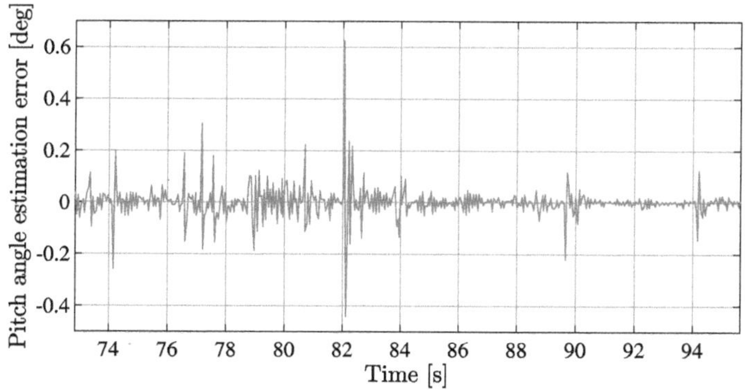

Figure 7.32 Pitch angle estimation error. Results of the experiment applying the Luenberger observer to a fixed-wing UAV.

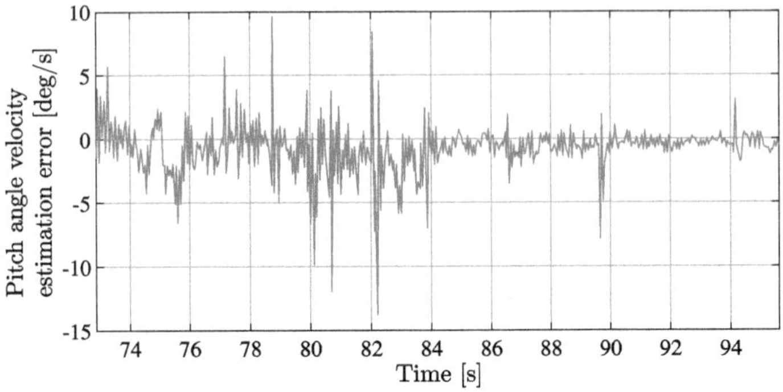

Figure 7.33 Pitch angle velocity estimation error. Results of the experiment applying the Luenberger observer to a fixed-wing UAV.

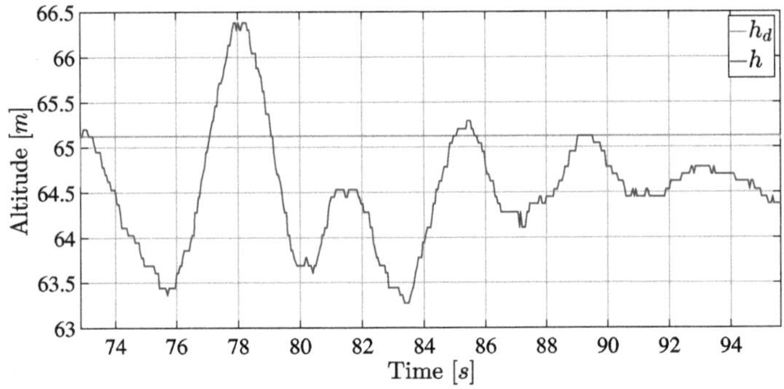

Figure 7.34 Measured altitude. Results of the experiment applying the Luenberger observer to a fixed-wing UAV.

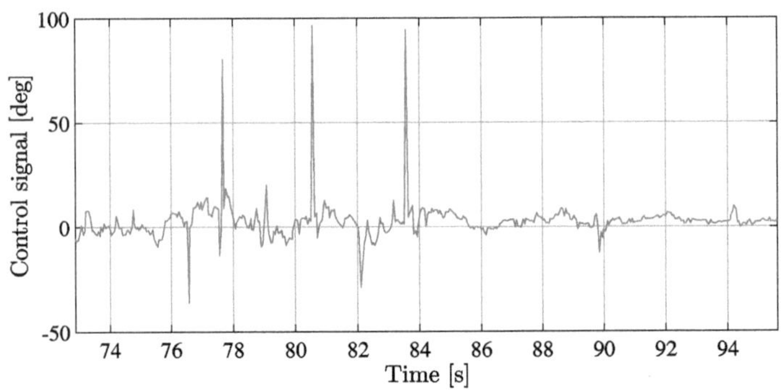

Figure 7.35 Control signal. Results of the experiment applying the Luenberger observer to a fixed-wing UAV.

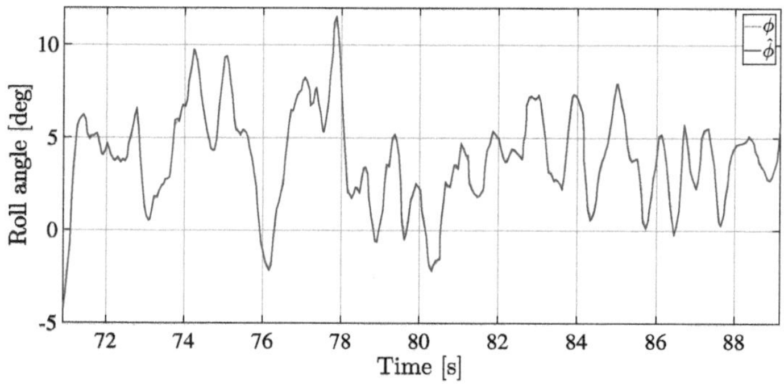

Figure 7.36 Results of the experiment applying the Luenberger observer to a fixed-wing UAV for roll angle.

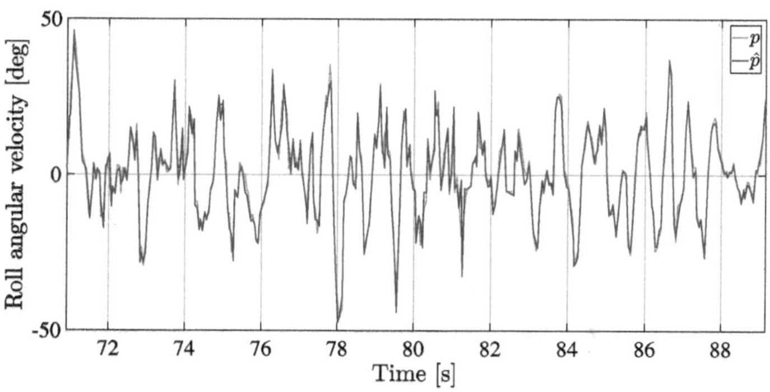

Figure 7.37 Results of the experiment applying the Luenberger observer to a fixed-wing UAV for roll angular velocity.

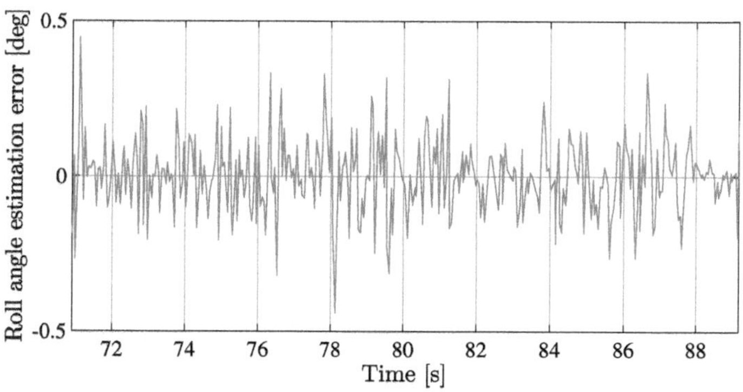

Figure 7.38 Roll angle estimation error. Results of the experiment applying the Luenberger observer to a fixed-wing UAV.

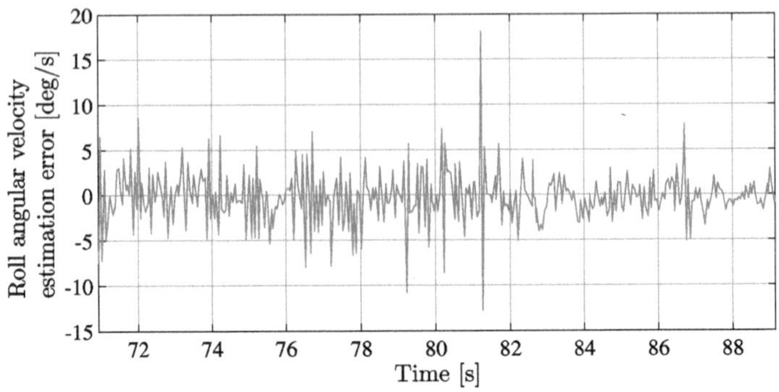

Figure 7.39 Roll angular velocity estimation error. Results of the experiment applying the Luenberger observer to a fixed-wing UAV.

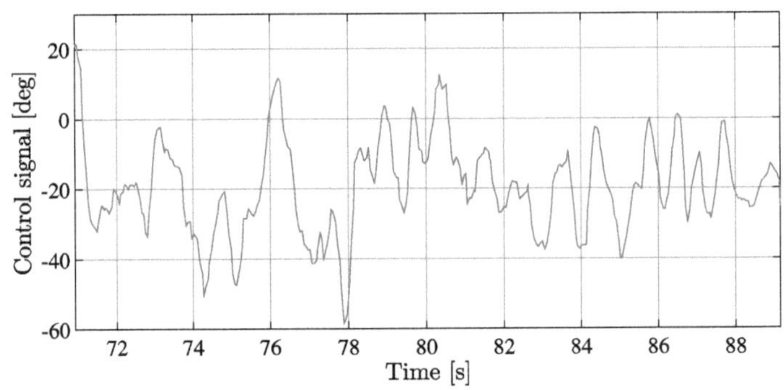

Figure 7.40 Control signal. Results of the experiment applying the Luenberger observer to a fixed-wing UAV for roll angle.

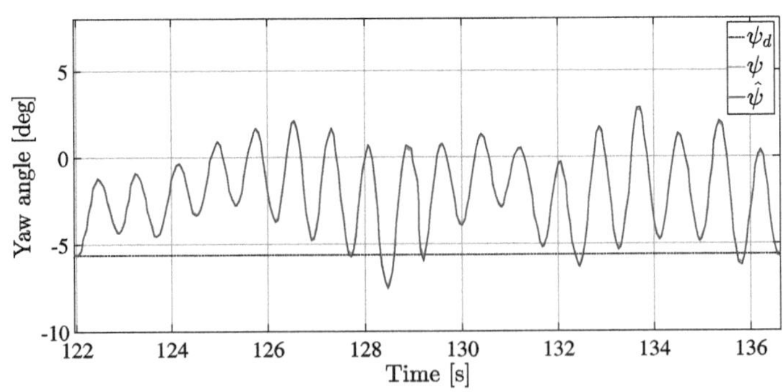

Figure 7.41 Results of the experiment applying the Luenberger observer to a fixed-wing UAV for yaw angle.

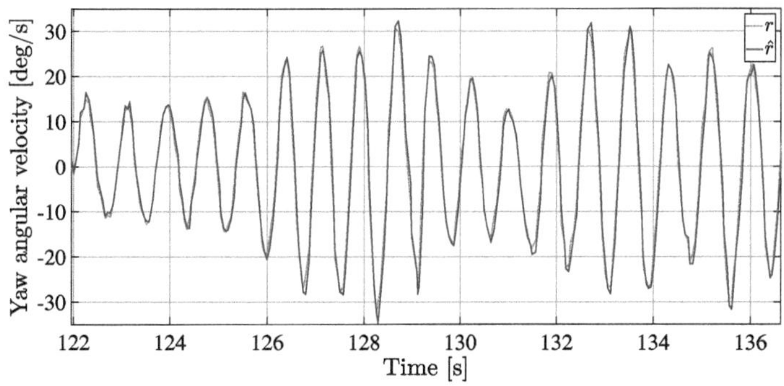

Figure 7.42 Results of the experiment applying the Luenberger observer to a fixed-wing UAV for yaw angular velocity.

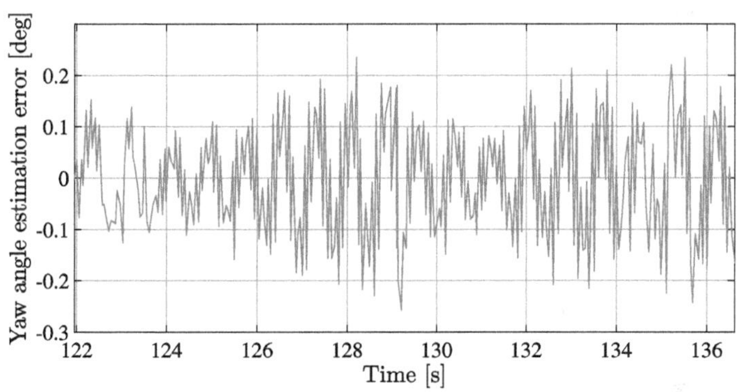

Figure 7.43 Yaw angle estimation error. Results of the experiment applying the Luenberger observer to a fixed-wing UAV.

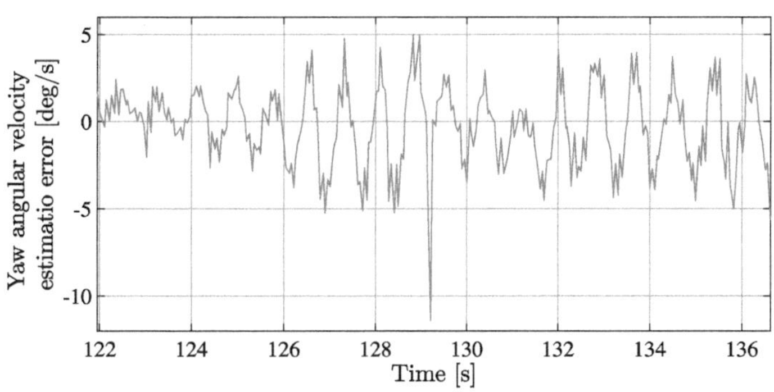

Figure 7.44 Estimation of the yaw angular velocity. Results of the experiment applying the Luenberger observer to a fixed-wing UAV.

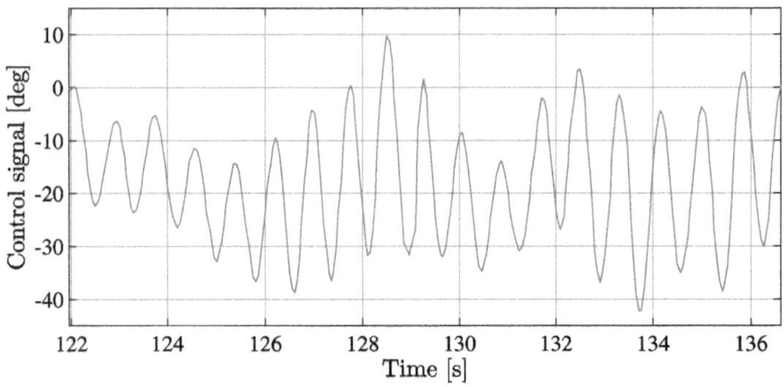

Figure 7.45 Control signal. Results of the experiment applying the Luenberger observer to a fixed-wing UAV for yaw angle.

7.10.2 SMO applied to a fixed-wing UAV with PD control law

The experimental results of SMO are presented for pitch, roll, yaw, and altitude movement of the UAV.

Pitch motion

Figs. 7.46 and 7.47 show the experimental results of SMO and a PD control law for pitch angle and pitch angular velocity, respectively. Figs. 7.48 and 7.49 show the angle estimation and angular velocity errors. Fig. 7.50 shows the experimental results of altitude control. Fig. 7.51 shows the control signal of the same experiment.

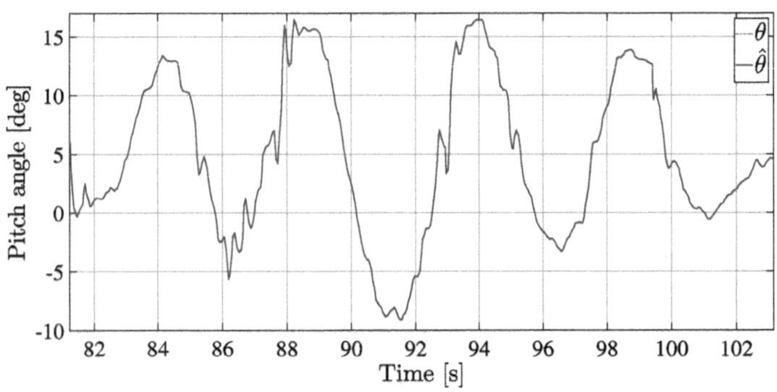

Figure 7.46 Results of the experiment applying SMO to a fixed-wing UAV for pitch angle.

Roll motion

Figs. 7.52 and 7.53 show the experimental results of SMO and a PD control law for the roll angle and roll angular velocity, respectively. Figs. 7.54 and 7.55 show the estimation angle and angular velocity errors. Fig. 7.56 shows the control signal from the same experiment.

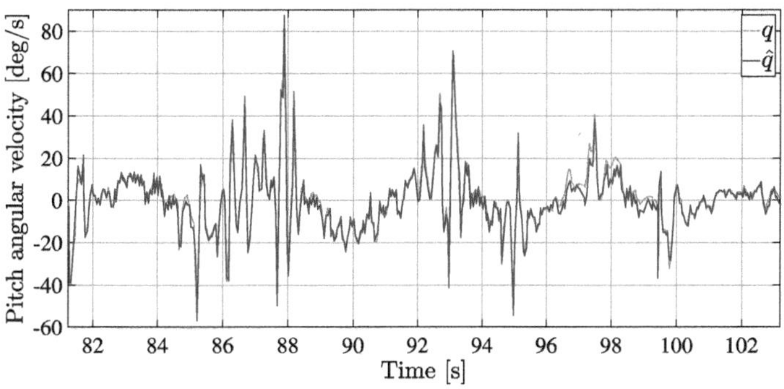

Figure 7.47 Results of the experiment applying SMO to a fixed-wing UAV for pitch angular velocity.

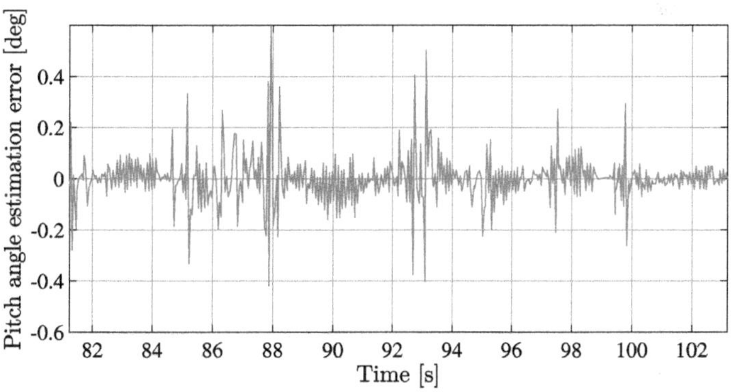

Figure 7.48 Pitch angle estimation error. Results of the experiment applying SMO to a fixed-wing UAV.

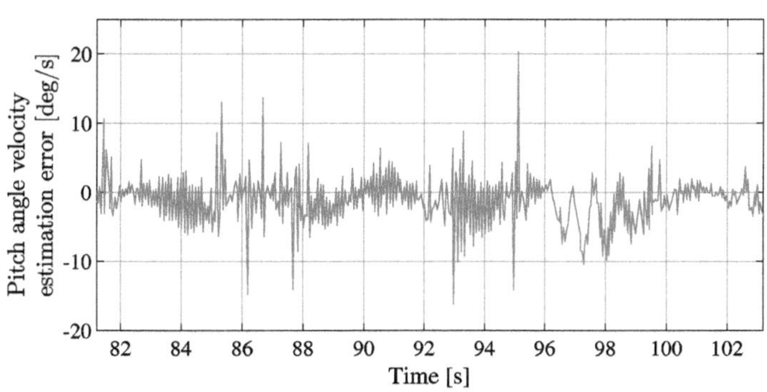

Figure 7.49 Pitch angle velocity estimation error. Results of the experiment applying SMO to a fixed-wing UAV.

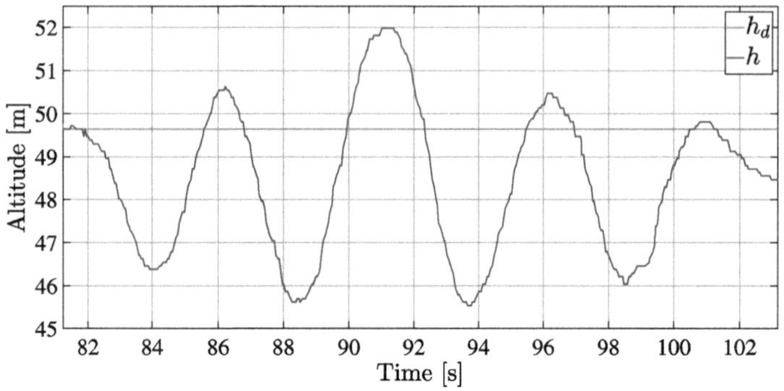

Figure 7.50 Measured altitude. Results of the experiment applying SMO to a fixed-wing UAV for pitch angle.

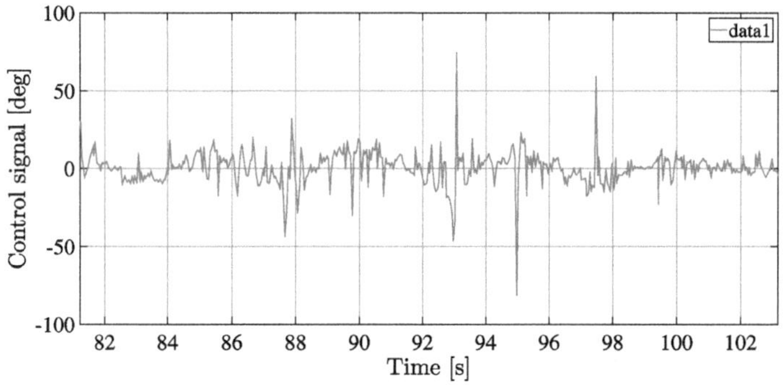

Figure 7.51 Control signal. Results of the experiment applying SMO to a fixed-wing UAV for pitch angle.

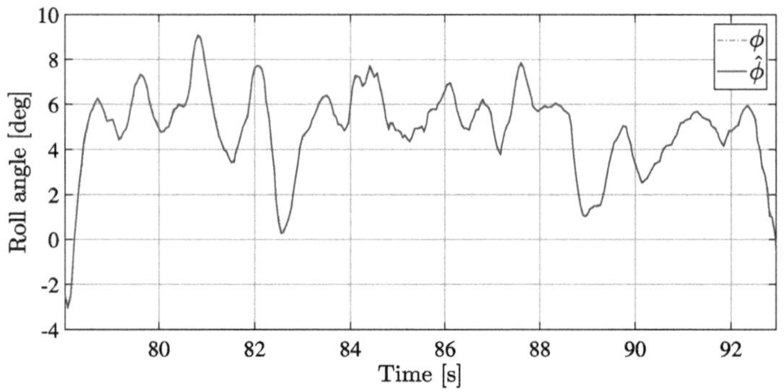

Figure 7.52 Results of the experiment applying SMO to a fixed-wing UAV for roll angle.

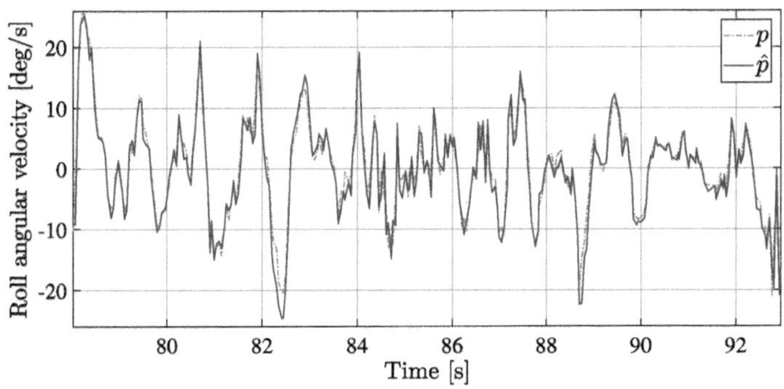

Figure 7.53 Results of the experiment applying SMO to a fixed-wing UAV for roll angular velocity.

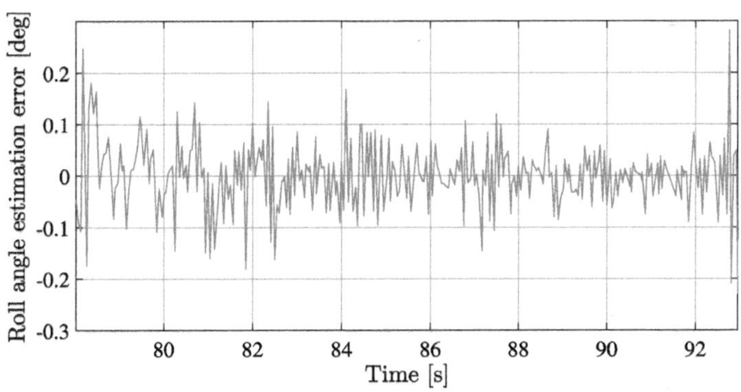

Figure 7.54 Roll angle estimation error. Results of the experiment applying SMO to a fixed-wing UAV.

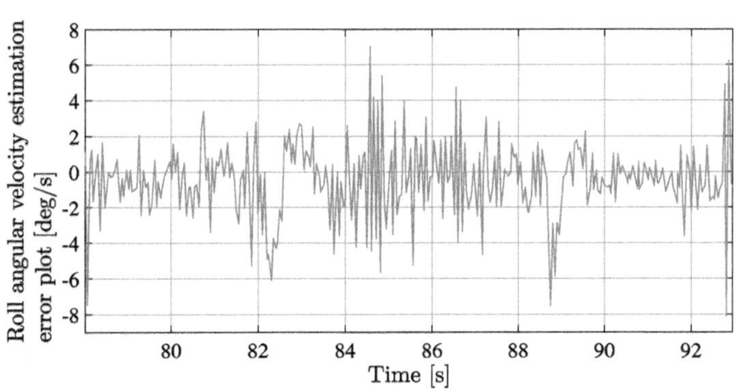

Figure 7.55 Roll angular velocity estimation error. Results of the experiment applying SMO to a fixed-wing UAV.

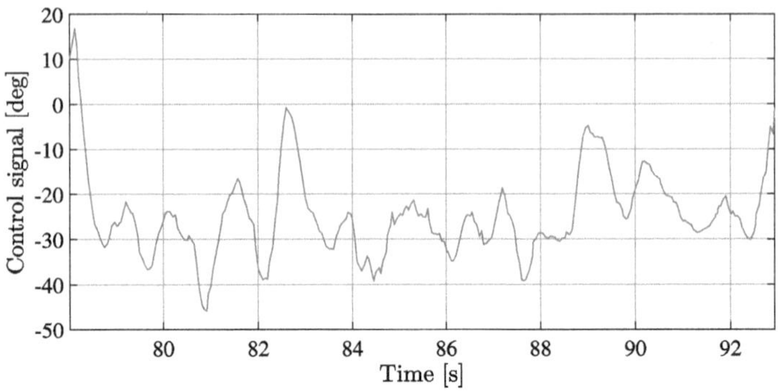

Figure 7.56 Control signal. Results of the experiment applying SMO to a fixed-wing UAV for roll angle.

Yaw motion

Figs. 7.57 and 7.58 show the experimental results of SMO and a PD control law for the yaw angle and yaw angular velocity, respectively. Figs. 7.59 and 7.60 show the angle estimation and angular velocity errors. Fig. 7.61 shows the control signal of the same experiment.

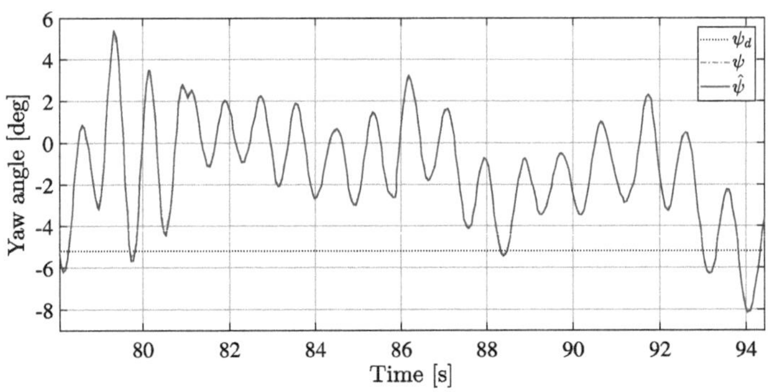

Figure 7.57 Results of the experiment applying SMO to a fixed-wing UAV for yaw angle.

7.10.3 NESO applied to a fixed-wing UAV with PD control law

The experimental results of NESO for pitch, roll, yaw, and altitude movement the UAV are presented.

Pitch motion

Figs. 7.62 and 7.63 show the experimental results of NESO and a PD control law for pitch angle and pitch angular velocity, respectively. Figs. 7.64 and 7.65 show the angle estimation and angular velocity errors. Fig. 7.66 shows experimental results of altitude control. Fig. 7.67 shows the control signal of the same experiment.

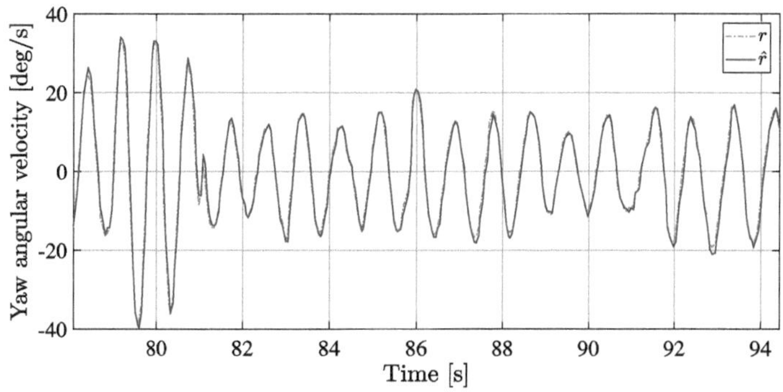

Figure 7.58 Results of the experiment applying SMO to a fixed-wing UAV for yaw angular velocity.

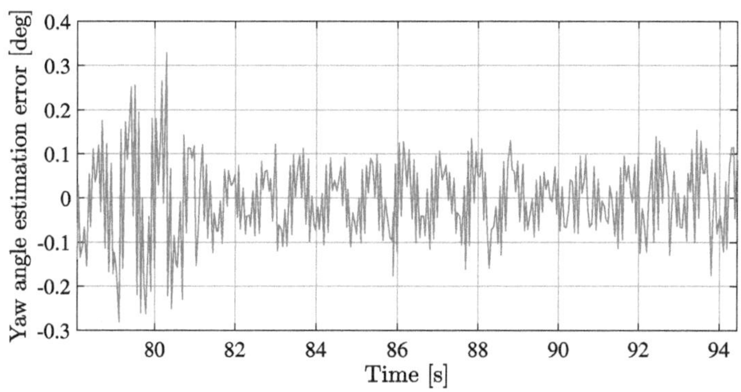

Figure 7.59 Yaw angle estimation error. Results of the experiment applying SMO to a fixed-wing UAV.

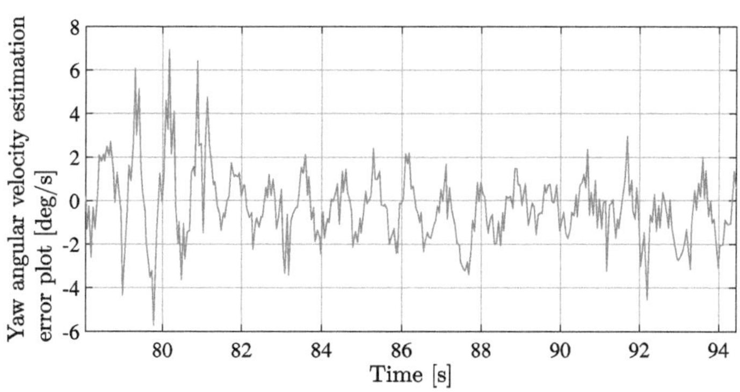

Figure 7.60 Estimation of the yaw angular velocity. Results of the experiment applying SMO to a fixed-wing UAV.

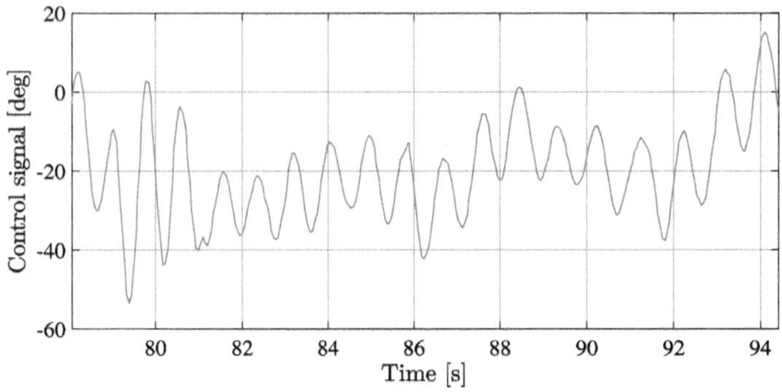

Figure 7.61 Control signal. Results of the experiment applying SMO to a fixed-wing UAV for yaw angle.

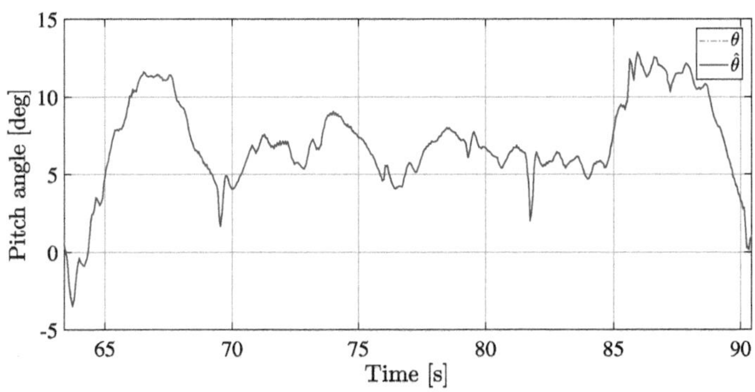

Figure 7.62 Results of the experiment applying NESO to a fixed-wing UAV for pitch angle.

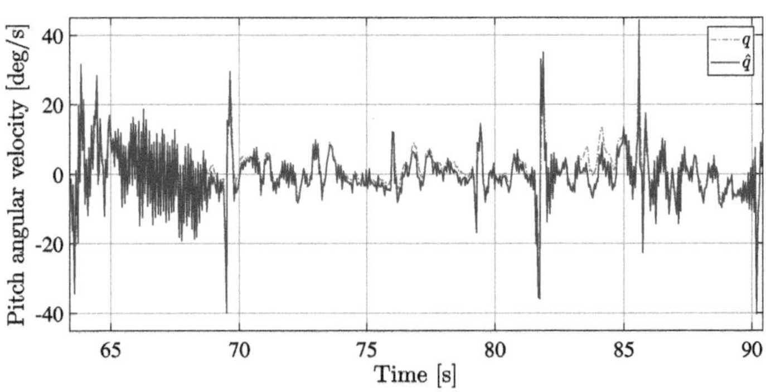

Figure 7.63 Results of the experiment applying NESO to a fixed-wing UAV for pitch angular velocity.

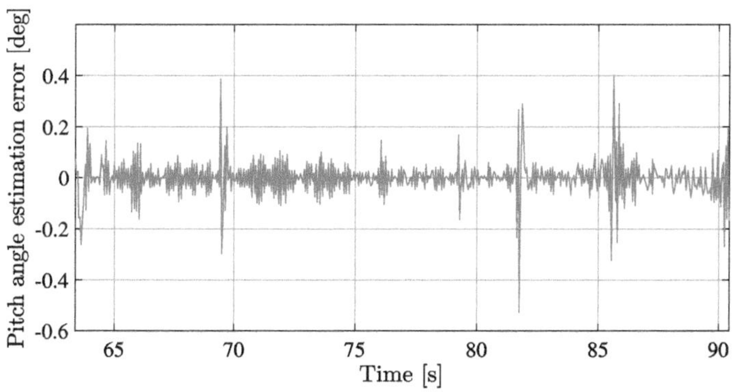

Figure 7.64 Pitch angle estimation error. Results of the experiment applying NESO to a fixed-wing UAV.

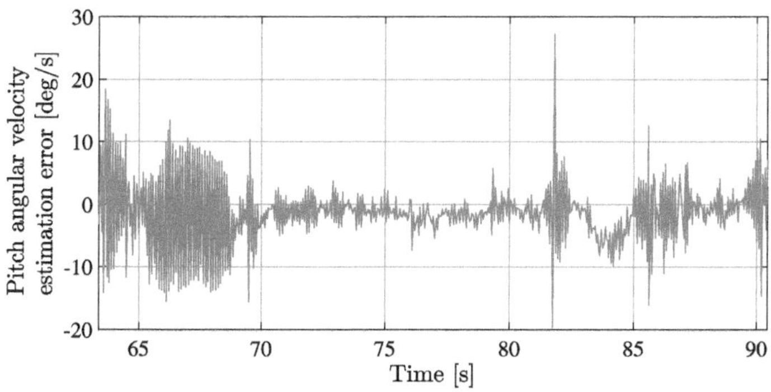

Figure 7.65 Pitch angle velocity estimation error. Results of the experiment applying NESO to a fixed-wing UAV.

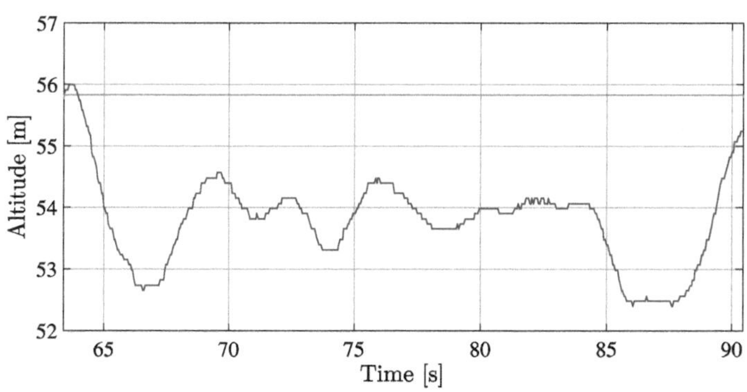

Figure 7.66 Measured altitude. Results of the experiment applying NESO to a fixed-wing UAV for pitch angle.

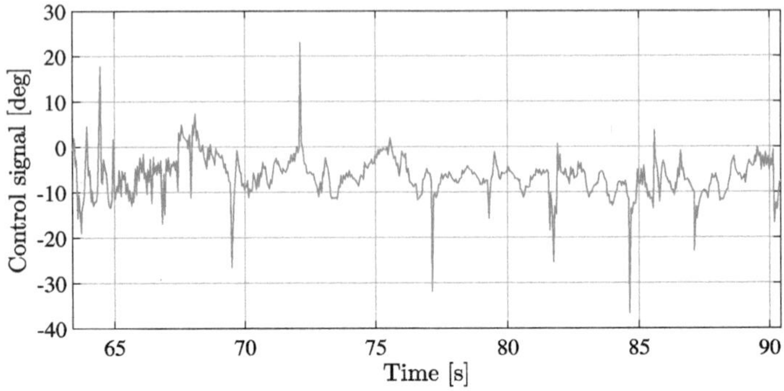

Figure 7.67 Control signal. Results of the experiment applying NESO to a fixed-wing UAV for pitch angle.

Roll motion

Figs. 7.68 and 7.69 show the experimental results of NESO and a PD control law for the roll angle and roll angular velocity, respectively. Figs. 7.70 and 7.71 show the estimation angle and angular velocity errors. Fig. 7.72 shows the control signal from the same experiment.

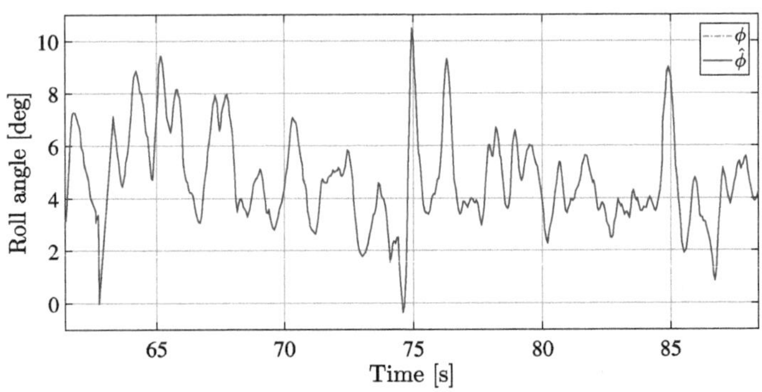

Figure 7.68 Results of the experiment applying NESO to a fixed-wing UAV for roll angle.

Yaw motion

Figs. 7.73 and 7.74 show the experimental results of NESO and a PD control law for the yaw angle and yaw angular velocity, respectively. Figs. 7.75 and 7.76 show the angle estimation and angular velocity errors. Fig. 7.77 shows the control signal of the same experiment.

7.10.4 SMO applied to a fixed-wing UAV with backstepping control law

Next, the experimental results of SMO with a backstepping control law are presented for pitch, roll, yaw, and altitude movement of the UAV.

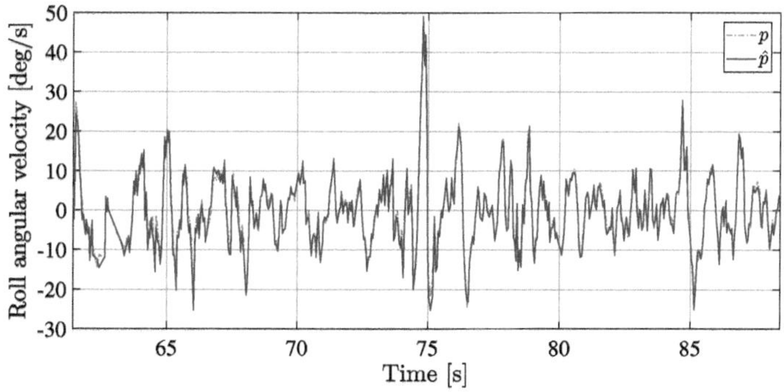

Figure 7.69 Results of the experiment applying NESO to a fixed-wing UAV for roll angular velocity.

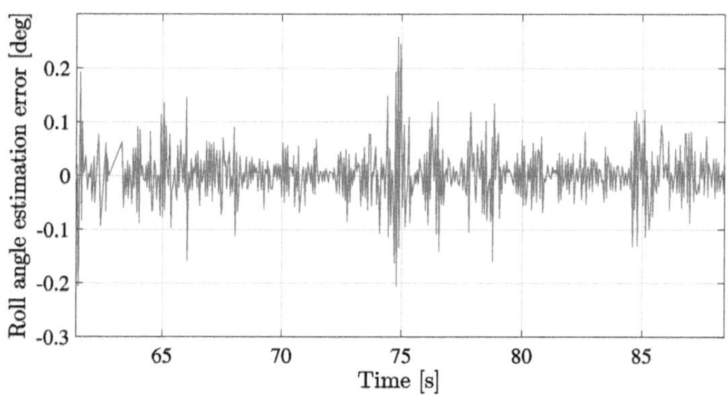

Figure 7.70 Roll angle estimation error. Results of the experiment applying NESO to a fixed-wing UAV.

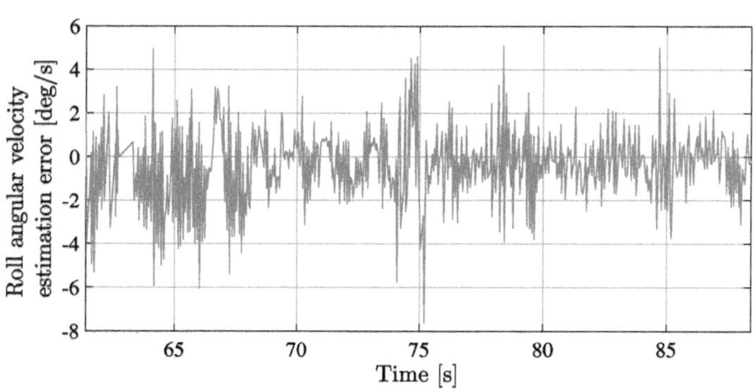

Figure 7.71 Roll angular velocity estimation error. Results of the experiment applying NESO to a fixed-wing UAV.

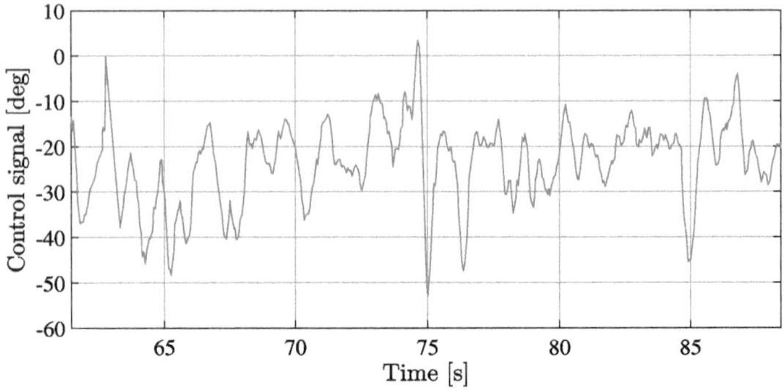

Figure 7.72 Control signal. Results of the experiment applying NESO to a fixed-wing UAV for roll angle.

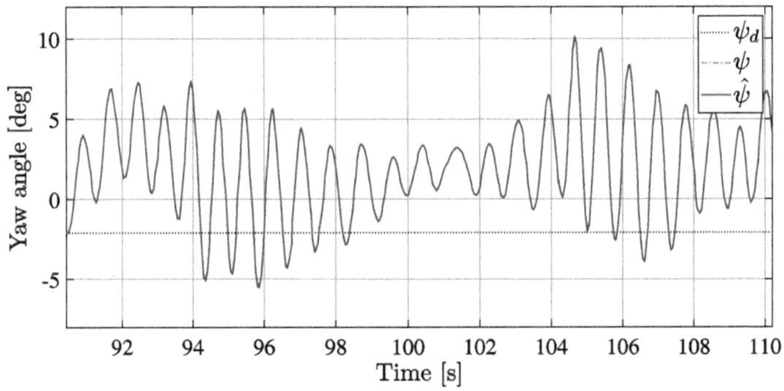

Figure 7.73 Results of the experiment applying NESO to a fixed-wing UAV for yaw angle.

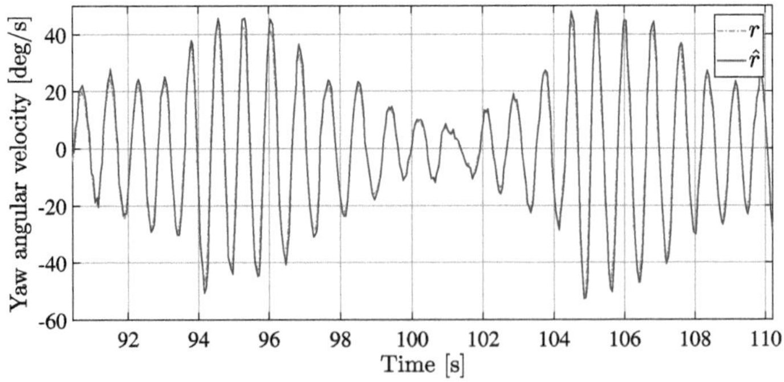

Figure 7.74 Results of the experiment applying NESO to a fixed-wing UAV for yaw angular velocity.

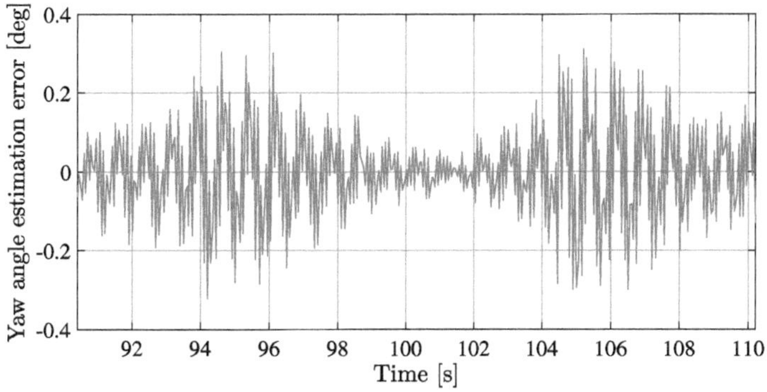

Figure 7.75 Yaw angle estimation error. Results of the experiment applying NESO to a fixed-wing UAV.

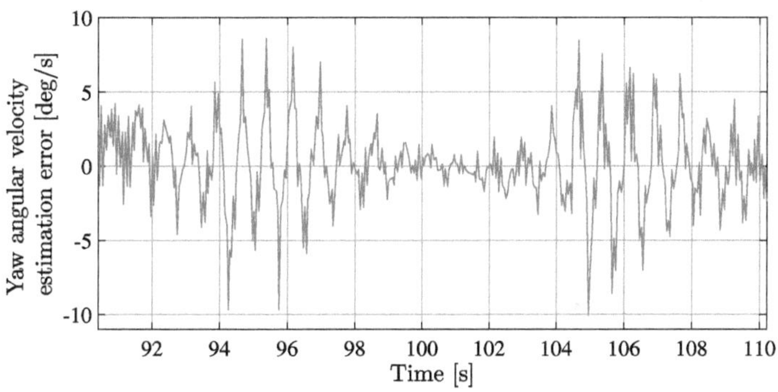

Figure 7.76 Estimation of the yaw angular velocity. Results of the experiment applying NESO to a fixed-wing UAV.

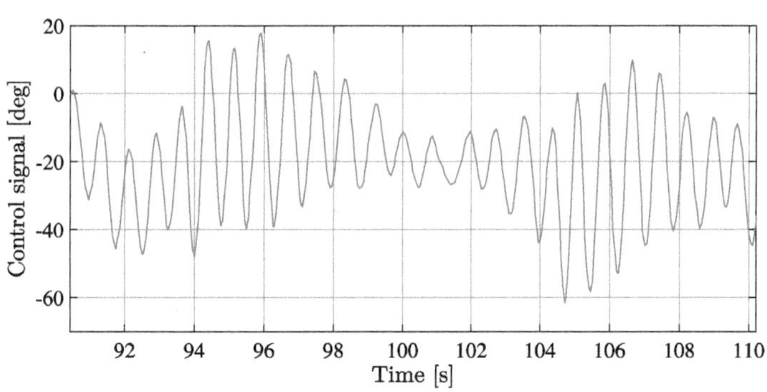

Figure 7.77 Control signal. Results of the experiment applying NESO to a fixed-wing UAV for yaw angle.

Pitch motion

Figs. 7.78 and 7.79 show the experimental results of SMO and a backstepping control law for pitch angle and pitch angular velocity, respectively. Figs. 7.80 and 7.81 show the angle estimation and angular velocity errors. Fig. 7.82 show experimental results of altitude control. Fig. 7.83 show the control signal of the same experiment.

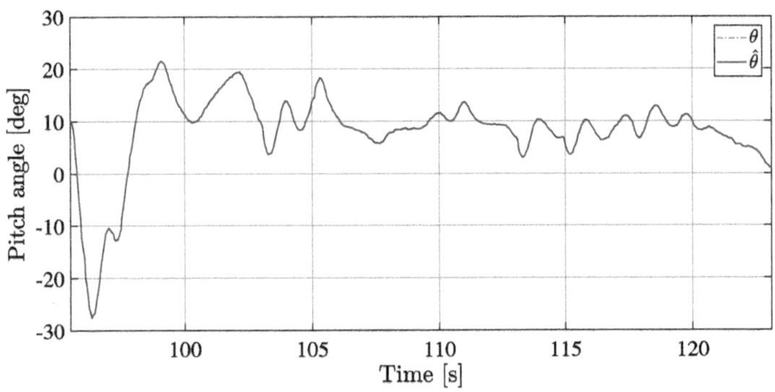

Figure 7.78 Results of the experiment applying SMO to a fixed-wing UAV for pitch angle.

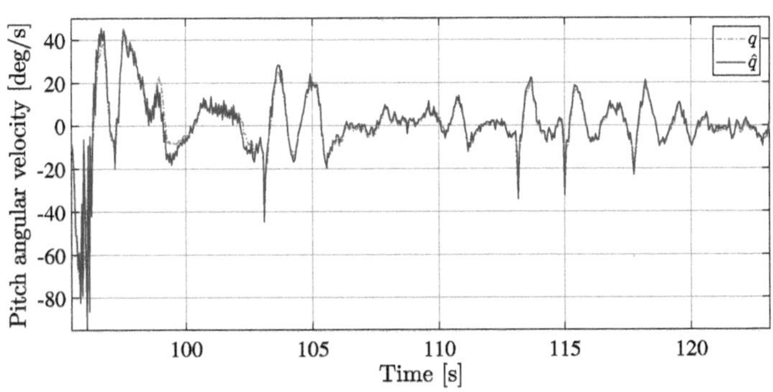

Figure 7.79 Results of the experiment applying SMO to a fixed-wing UAV for pitch angular velocity.

Roll motion

Figs. 7.84 and 7.85 show the experimental results of SMO and a backstepping control law for the roll angle and roll angular velocity, respectively. Figs. 7.86 and 7.87 show the estimation angle and angular velocity errors. Fig. 7.88 shows the control signal from the same experiment.

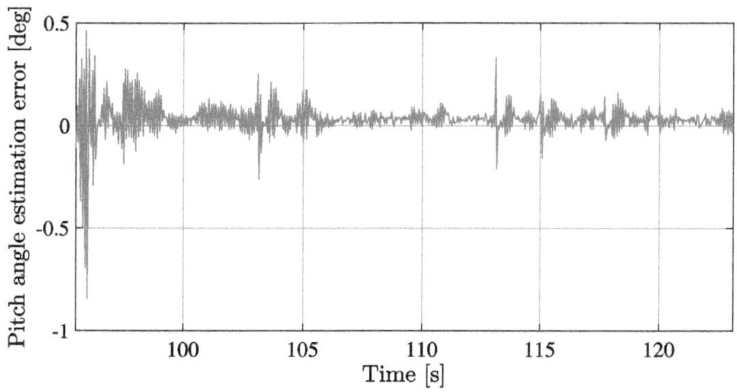

Figure 7.80 Pitch angle estimation error. Results of the experiment applying SMO to a fixed-wing UAV.

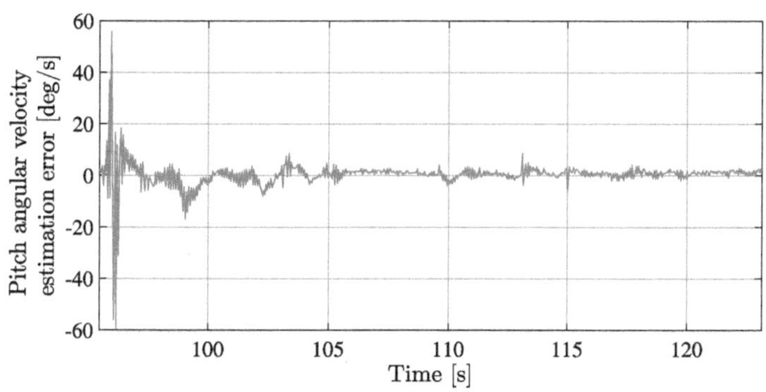

Figure 7.81 Pitch angle velocity estimation error. Results of the experiment applying SMO to a fixed-wing UAV.

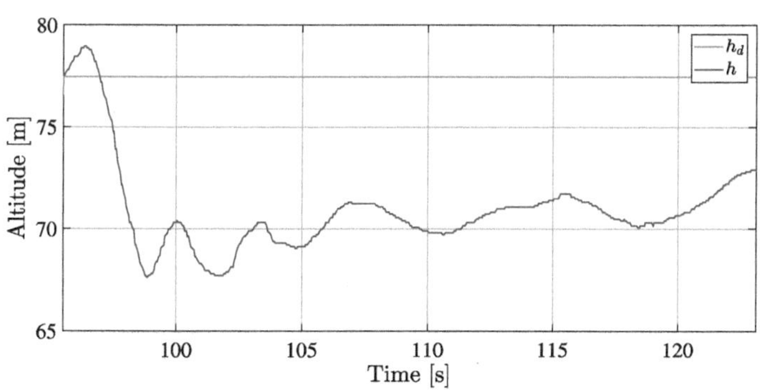

Figure 7.82 Measured altitude. Results of the experiment applying SMO to a fixed-wing UAV for pitch angle.

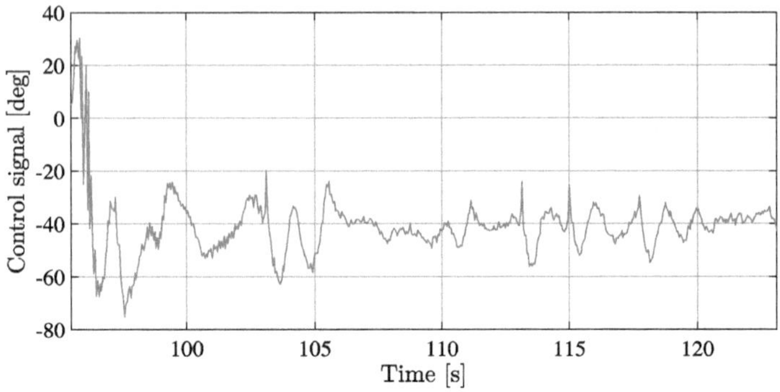

Figure 7.83 Control signal. Results of the experiment applying SMO to a fixed-wing UAV for pitch angle.

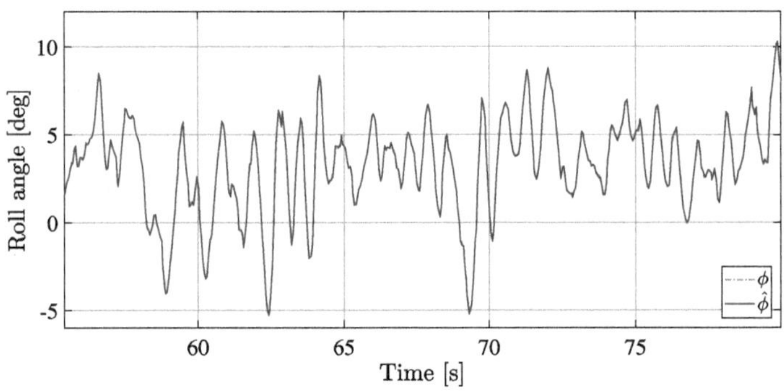

Figure 7.84 Results of the experiment applying SMO to a fixed-wing UAV for roll angle.

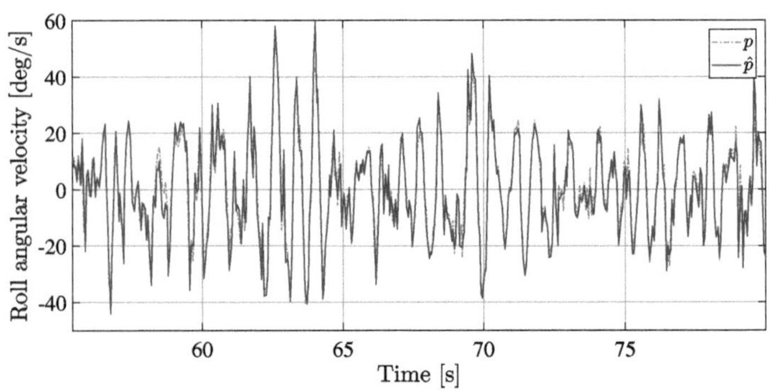

Figure 7.85 Results of the experiment applying SMO to a fixed-wing UAV for roll angular velocity.

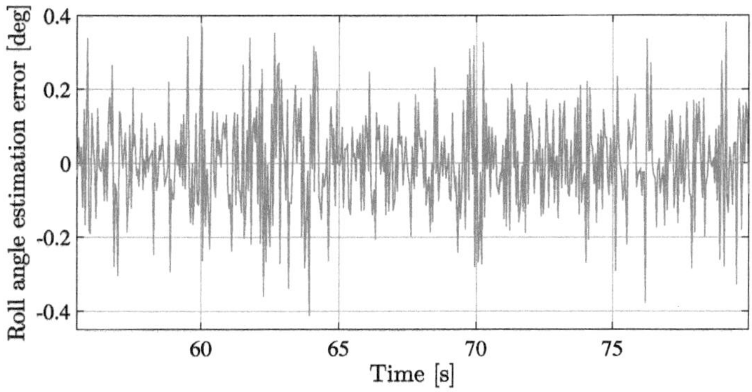

Figure 7.86 Roll angle estimation error. Results of the experiment applying SMO to a fixed-wing UAV.

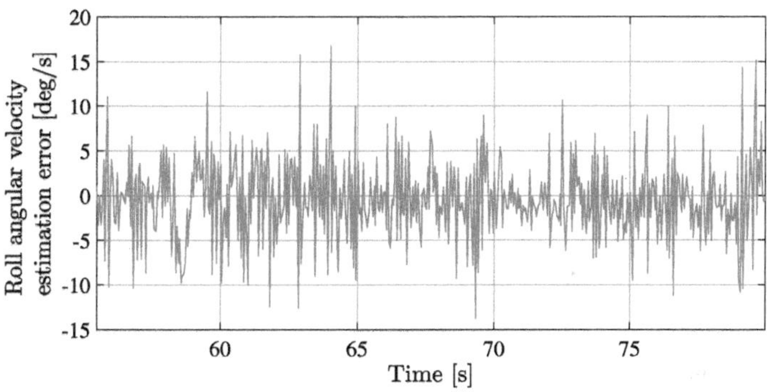

Figure 7.87 Roll angular velocity estimation error. Results of the experiment applying SMO to a fixed-wing UAV.

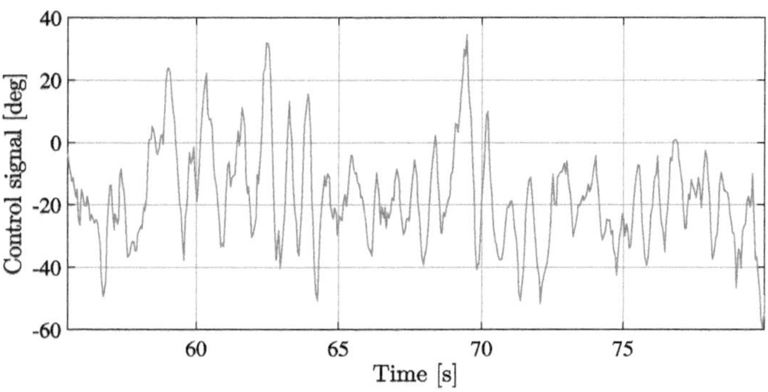

Figure 7.88 Control signal. Results of the experiment applying SMO to a fixed-wing UAV for roll angle.

Yaw motion

Figs. 7.89 and 7.90 show the experimental results of SMO and a backstepping control law for the yaw angle and yaw angular velocity, respectively. Figs. 7.91 and 7.92 show the angle estimation and angular velocity errors. Fig. 7.93 shows the control signal of the same experiment.

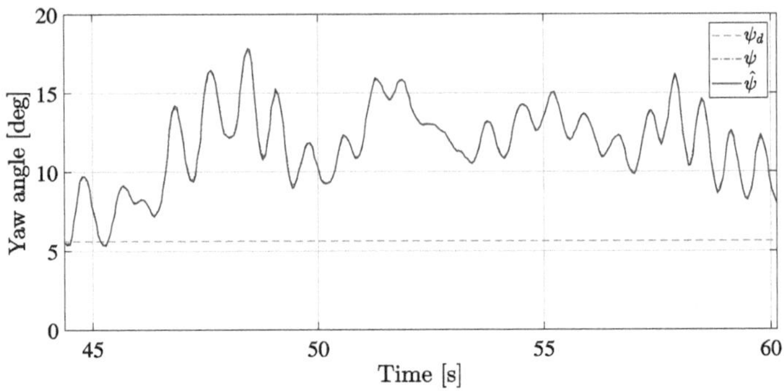

Figure 7.89 Results of the experiment applying SMO to a fixed-wing UAV for yaw angle.

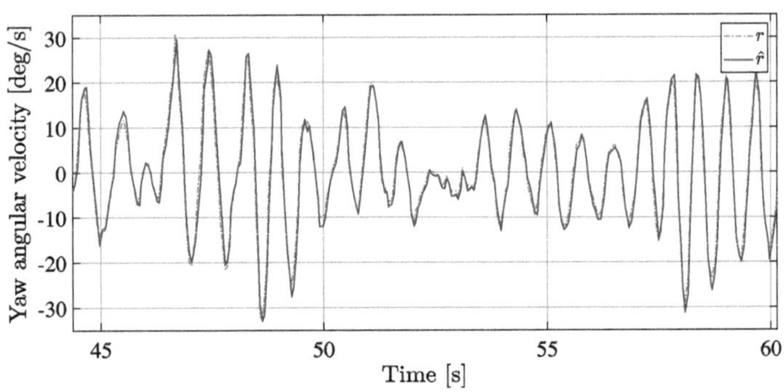

Figure 7.90 Results of the experiment applying SMO to a fixed-wing UAV for yaw angular velocity.

7.10.5 NESO applied to a fixed-wing UAV with backstepping control law

The results of the experiments applying NESO and a backstepping control law for pitch, roll, yaw, and altitude movement of the UAV are presented.

Pitch motion

Figs. 7.94 and 7.95 show the experimental results of NESO and a backstepping control law for pitch angle and pitch angular velocity, respectively. Figs. 7.96 and 7.97 show

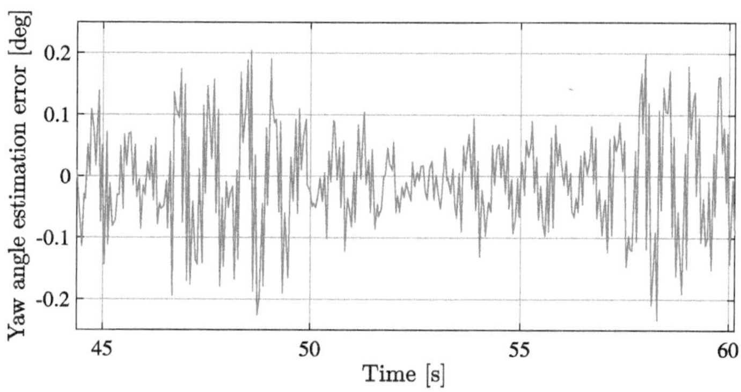

Figure 7.91 Yaw angle estimation error. Results of the experiment applying SMO to a fixed-wing UAV.

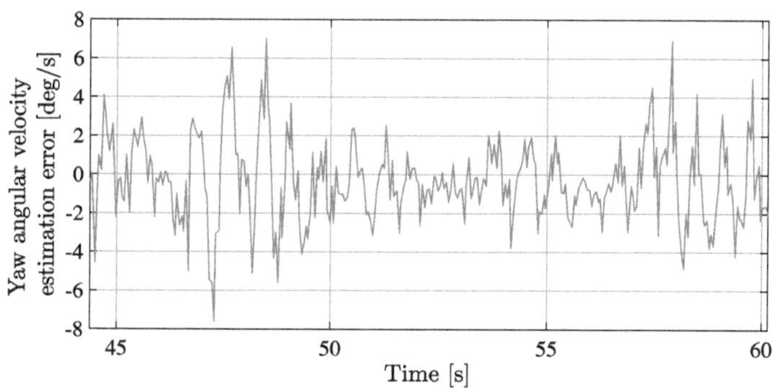

Figure 7.92 Estimation of the yaw angular velocity. Results of the experiment applying SMO to a fixed-wing UAV.

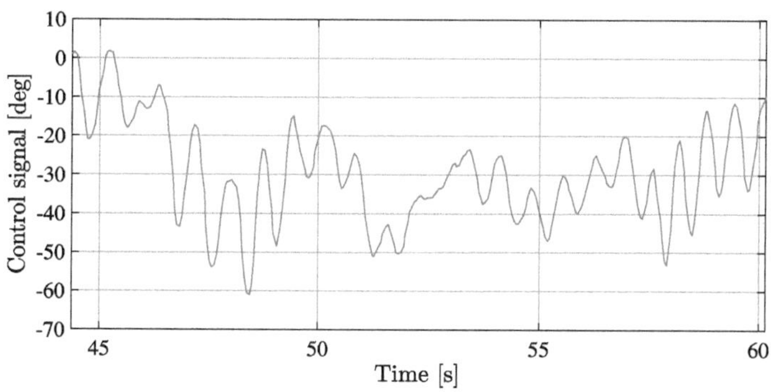

Figure 7.93 Control signal. Results of the experiment applying SMO to a fixed-wing UAV for yaw angle.

the angle estimation and angular velocity errors. Fig. 7.98 shows experimental results of altitude control. Fig. 7.99 shows the control signal of the same experiment.

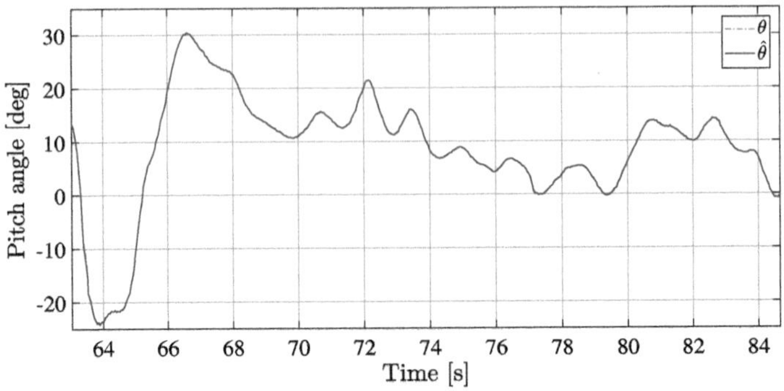

Figure 7.94 Results of the experiment applying NESO to a fixed-wing UAV for pitch angle.

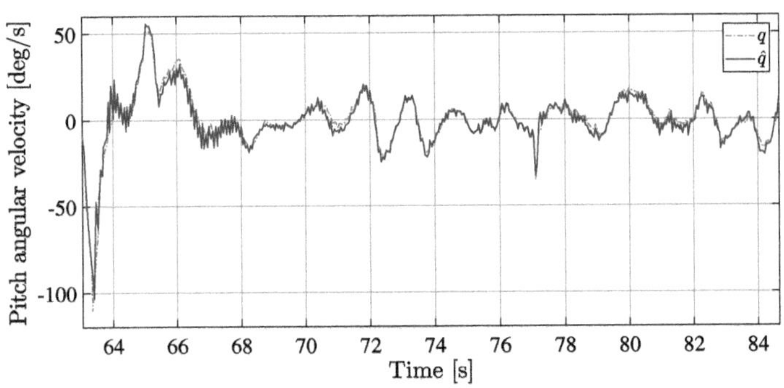

Figure 7.95 Results of the experiment applying NESO to a fixed-wing UAV for pitch angular velocity.

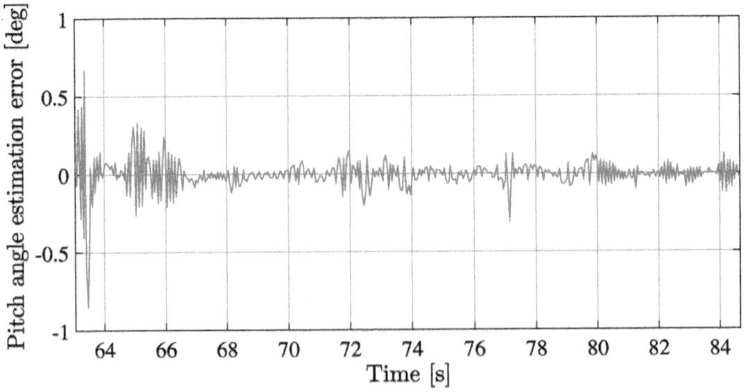

Figure 7.96 Pitch angle estimation error. Results of the experiment applying NESO to a fixed-wing UAV.

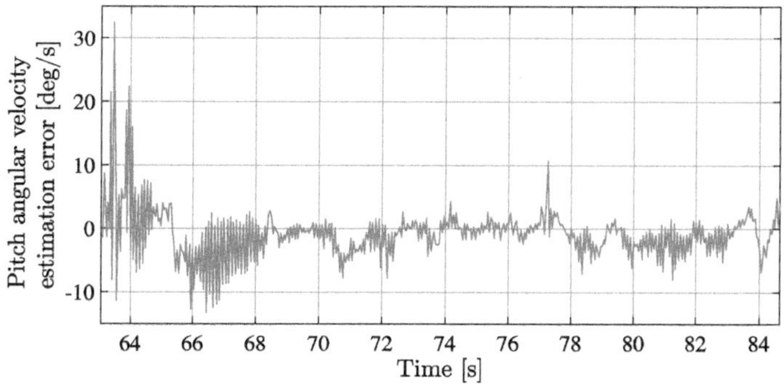

Figure 7.97 Pitch angle velocity estimation error. Results of the experiment applying NESO to a fixed-wing UAV.

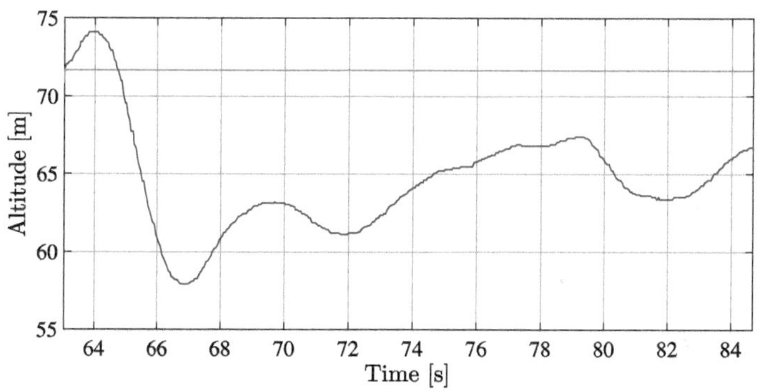

Figure 7.98 Measured altitude. Results of the experiment applying NESO to a fixed-wing UAV for pitch angle.

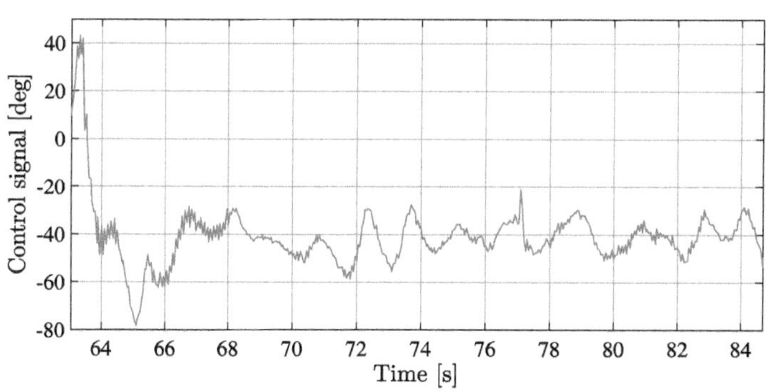

Figure 7.99 Control signal. Results of the experiment applying NESO to a fixed-wing UAV for pitch angle.

Roll motion

Figs. 7.100 and 7.101 show the experimental results of NESO and a backstepping control law for the roll angle and roll angular velocity, respectively. Figs. 7.102 and 7.103 show the estimation angle and angular velocity errors. Fig. 7.104 shows the control signal from the same experiment.

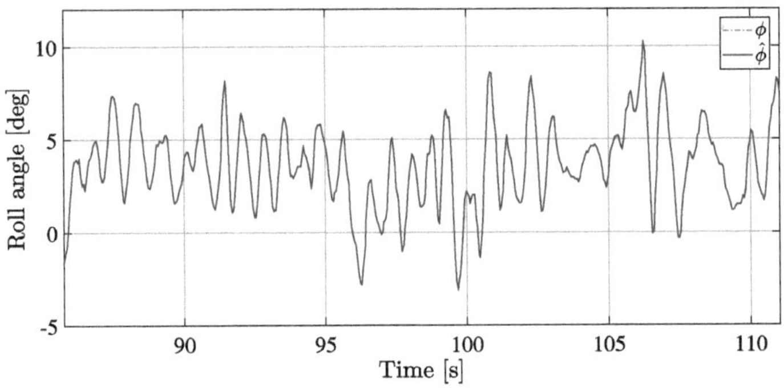

Figure 7.100 Results of the experiment applying NESO to a fixed-wing UAV for roll angle.

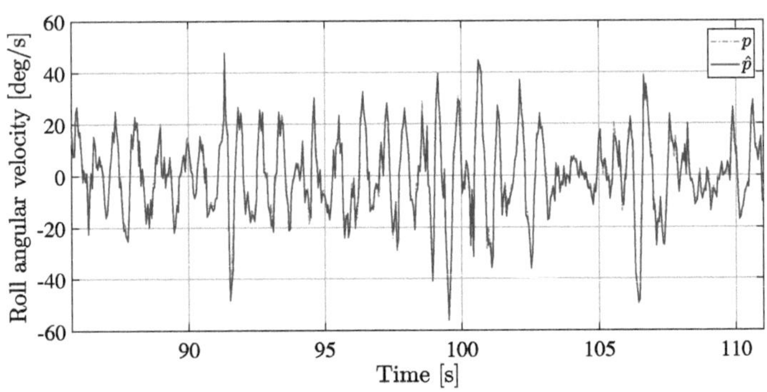

Figure 7.101 Results of the experiment applying NESO to a fixed-wing UAV for roll angular velocity.

Yaw motion

Figs. 7.105 and 7.106 show the experimental results of NESO and a backstepping control law for the yaw angle and yaw angular velocity, respectively. Figs. 7.107 and 7.108 show the angle estimation and angular velocity errors. Fig. 7.109 shows the control signal of the same experiment.

Table 7.6 shows the calculated norms of the estimation errors for the two control laws and the state observers. These norms are calculated with the following function:

$$\|\tilde{x}_i\| = \|x_i - \hat{x}_i\| = \sqrt{\frac{1}{T - t_0} \int_{t_0}^{T} \left(x_i - \hat{x}_i\right)^2 dt}. \tag{7.2}$$

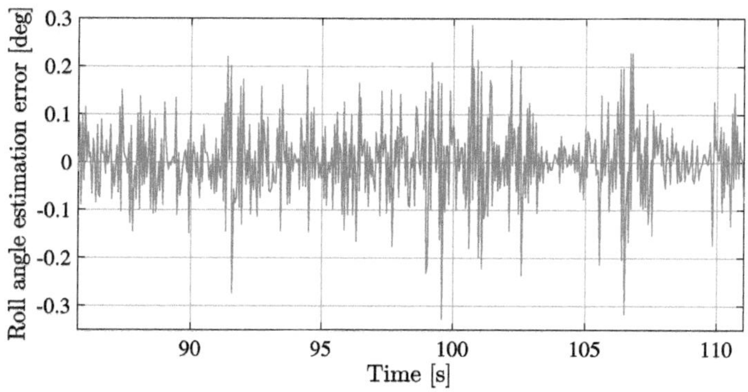

Figure 7.102 Roll angle estimation error. Results of the experiment applying NESO to a fixed-wing UAV.

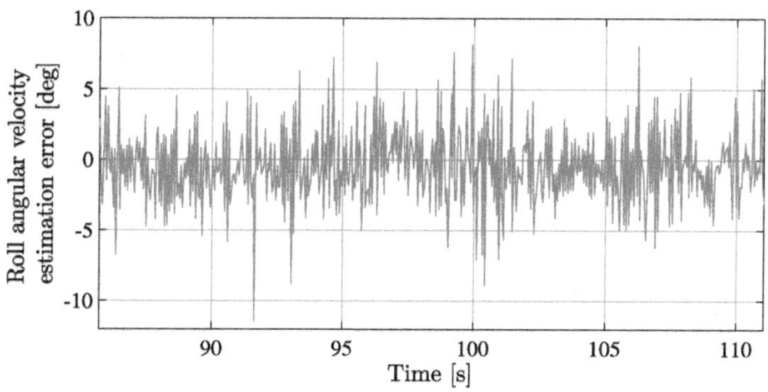

Figure 7.103 Roll angular velocity estimation error. Results of the experiment applying NESO to a fixed-wing UAV.

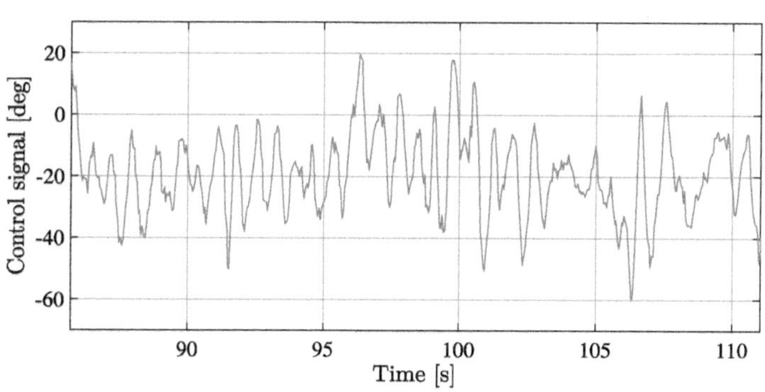

Figure 7.104 Control signal. Results of the experiment applying NESO to a fixed-wing UAV for roll angle.

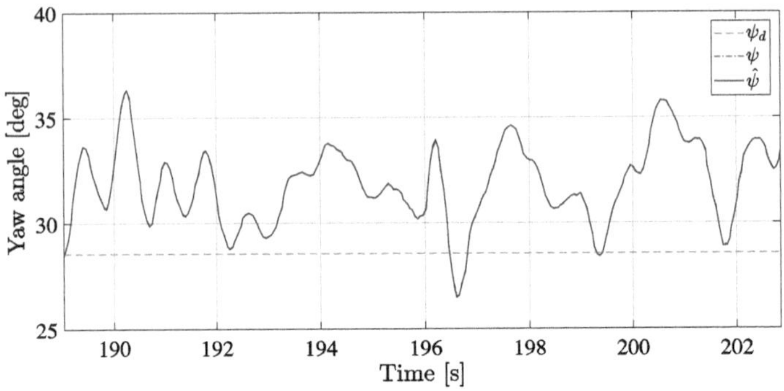

Figure 7.105 Results of the experiment applying NESO to a fixed-wing UAV for yaw angle.

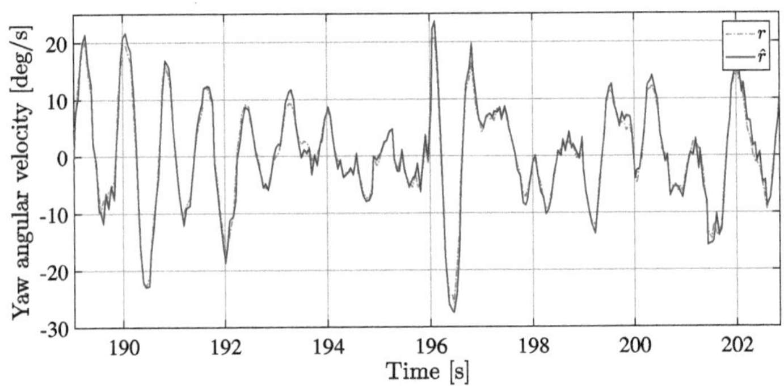

Figure 7.106 Results of the experiment applying NESO to a fixed-wing UAV for yaw angular velocity.

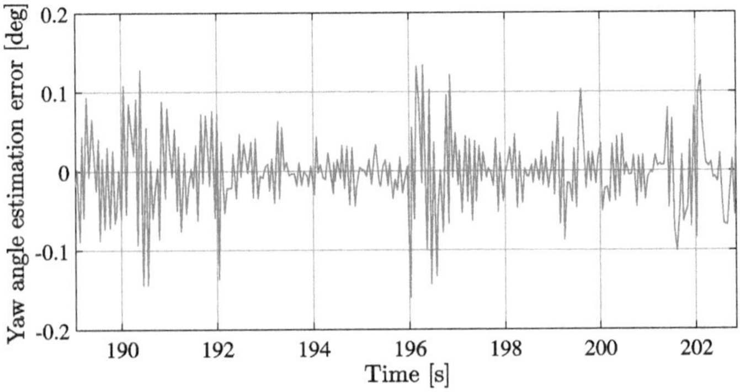

Figure 7.107 Yaw angle estimation error. Results of the experiment applying NESO to a fixed-wing UAV.

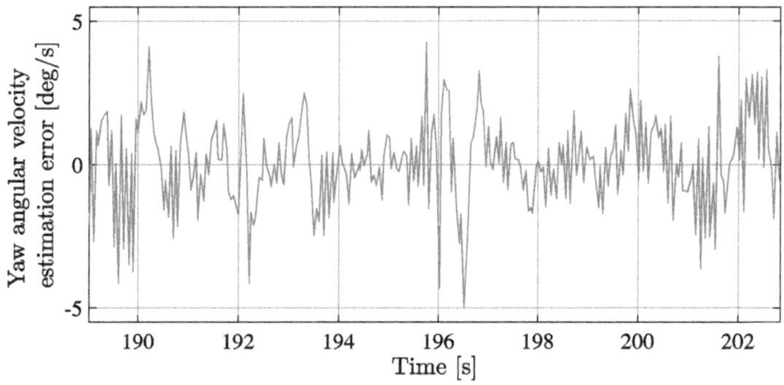

Figure 7.108 Estimation of the yaw angular velocity. Results of the experiment applying NESO to a fixed-wing UAV.

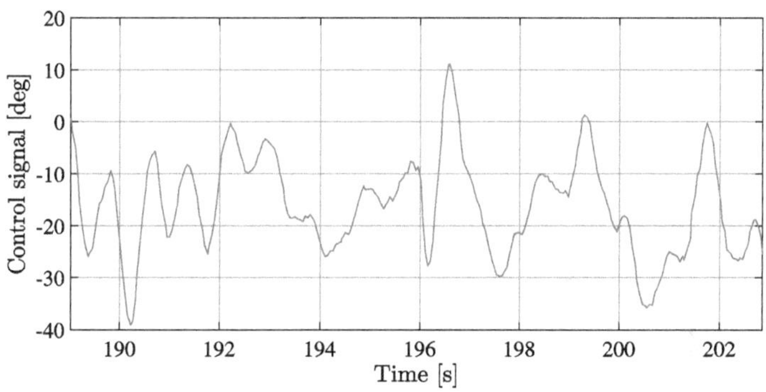

Figure 7.109 Control signal. Results of the experiment applying NESO to a fixed-wing UAV for yaw angle.

Table 7.6 $\mathcal{L}_2$ norm for the errors of the observers with different control laws.

Controller	Observer	Pitch [deg]		Yaw [deg]		Roll [deg]	
		$\mathcal{L}_2[\tilde{\theta}]$	$\mathcal{L}_2[\tilde{q}]$	$\mathcal{L}_2[\tilde{\phi}]$	$\mathcal{L}_2[\tilde{p}]$	$\mathcal{L}_2[\tilde{\psi}]$	$\mathcal{L}_2[\tilde{r}]$
PD	Luenberger	0.0638	2.1592	0.1183	2.9535	0.1102	2.2610
PD	SMO	0.0977	3.8484	0.0650	2.1717	0.0954	1.6894
PD	NESO	0.0977	3.8484	0.0650	2.1717	0.0954	1.6894
Backstepping	SMO	0.0971	5.6682	0.1364	4.5663	0.0853	2.1451
Backstepping	NESO	0.1049	4.3003	0.0858	2.7920	0.0492	1.5168

It was calculated for each movement. In the case of the yaw motion, x_1 and x_2 refer to $x_1 = \psi$ and $x_2 = r$, i.e., the yaw angle and the yaw rate, respectively, so Eq. (7.2) is defined for $\|\psi - \hat{\psi}\|$ and $\|r - \hat{r}\|$. The same analysis was done for pitch and roll movements.

7.11 Concluding remarks

This chapter described the testbed developed to implement the linear and nonlinear control techniques and the state observers in fixed-wing UAVs. We have presented the electronics used in the embedded system to develop an autopilot capable of supporting linear and nonlinear controllers and observers in software and hardware. Experimental results have been presented of some linear and nonlinear control techniques and state observers applied to fixed-wing UAVs.

Mathematical review

A

This appendix presents a basic review of mathematical fundamentals which are important to understand the theory presented in this book.

A.1 Vectors

A vector is a set of n values arranged in column form, that is,

$$x = \begin{bmatrix} x_1 \\ x_2 \\ \vdots \\ x_n \end{bmatrix} = \begin{bmatrix} x_1 & x_2 & \dots & x_n \end{bmatrix}^T, \tag{A.1}$$

where $x_1, x_2, \dots, x_n$ are the coordinates or components of the vector x. The superscript ^{T} represents the transpose of the vector x.

The set of all vectors of dimension n is called *Euclidean space* of dimension n and is denoted as $\mathbb{R}^n$. So, if x is a vector of dimension n, it can be represented as $x \in \mathbb{R}^n$. Euclidean space of dimension one is equivalent to the set of all real numbers $\mathbb{R}$.

Vectors can be added by simply adding their components. The only condition is that they are of the same dimension. Let $x, y, z \in \mathbb{R}^n$. Then

$$z = x + y = \begin{bmatrix} x_1 \\ x_2 \\ \vdots \\ x_n \end{bmatrix} + \begin{bmatrix} y_1 \\ y_2 \\ \vdots \\ y_n \end{bmatrix} = \begin{bmatrix} x_1 + y_1 \\ x_2 + y_2 \\ \vdots \\ x_n + y_n \end{bmatrix} = \begin{bmatrix} z_1 \\ z_2 \\ \vdots \\ z_n \end{bmatrix}. \tag{A.2}$$

If a vector $x \in \mathbb{R}^n$ is multiplied by a value scalar α, then each element of x is multiplied by α. On the other hand, the *inner product* of two vectors $x, y \in \mathbb{R}^n$ is defined as

$$x^T y = \begin{bmatrix} x_1 & x_2 & \dots & x_n \end{bmatrix} \begin{bmatrix} y_1 \\ y_2 \\ \vdots \\ y_n \end{bmatrix} = x_1 y_1 + x_2 y_2 + \dots + x_n y_n = \sum_{i=1}^{n} x_i y_i. \tag{A.3}$$

If the inner product is equal to zero, the two vectors are said to be orthogonal or perpendicular. It can be verified that the inner product has the following two properties:

- $x^T y = y^T x$, for all $x, y \in \mathbb{R}^n$,
- $x^T (y + z) = x^T y + x^T z$, for all $x, y, z \in \mathbb{R}^n$.

A.2 Linear transformations

This section presents the vectorial functions of vectorial variables, that is, functions with the form $w = F(v)$, where the dependent variable w and the independent variable v are vectors. Then, with the previous considerations, the study of a particular class of functions called linear transformations is presented. If V and W are vectorial spaces and F is a function which associates every vector in V with a single vector in W, then it is said that F *maps* V in W, and this is denoted as $F = V \longrightarrow W$. Also, if F associates the vector w to vector v, it is denoted as $w = F(v)$ and w is the *image* of v subjected to F. Then, by definition, if $F = V \longrightarrow W$ is a function that maps the vector space V to the vector space W, it is said that F is a linear transformation if:

- $F(u+v) = F(u) + F(v)$ for all vectors u and v in V,
- $F(ku) = kF(u)$ for all vectors u in V and for all scalars.

A.3 Euclidean norm

The *Euclidean norm* of a vector $x \in \mathbb{R}^n$, denoted as $\|x\|$, is defined as

$$\|x\| = \sqrt{\sum_{i=1}^{n} x_i^2} = \sqrt{x^T x}, \tag{A.4}$$

of which only the positive part of the square root is considered. Also, you can easily check the following properties:

- $\|x\| = 0$, if and only if $x = 0 \in \mathbb{R}^n$,
- $\|x\| > 0$, for all $x \in \mathbb{R}^n$ with $x \neq 0 \in \mathbb{R}^n$,
- $\|\alpha x\| = |\alpha| \|x\|$, for all $\alpha \in \mathbb{R}$ and $x \in \mathbb{R}^n$,
- $\|x\| - \|y\| \leq \|x + y\| \leq \|x\| + \|y\|$, for all $x, y \in \mathbb{R}^n$ (*triangle inequality*),
- $|x^T y| \leq \|x\| \|y\|$, for all $x, y \in \mathbb{R}^n$ (*Schwarz inequality*).

A.4 Matrices

Matrices are arrays of real values that are formed by n rows and m columns, which are called arrays of dimension $n \times m$:

$$A = \begin{bmatrix} a_{11} & a_{12} & \dots & a_{1m} \\ a_{21} & a_{22} & \dots & a_{2m} \\ \vdots & \vdots & \ddots & \vdots \\ a_{n1} & a_{n2} & \dots & a_{nm} \end{bmatrix}, \tag{A.5}$$

where $a_{ij} \in \mathbb{R}$, $i = 1, \ldots, n$, $j = 1, \ldots, m$, are the elements of the array A. The set of matrices of dimension $n \times m$ will be denoted by $\mathbb{R}^{n\times m}$, so $A \in \mathbb{R}^{n\times m}$. Then, a vector $x \in \mathbb{R}^n$ can be interpreted as a particular matrix belonging to $\mathbb{R}^{n\times 1} = \mathbb{R}^n$.

Let $A \in \mathbb{R}^{n\times m}$ be a matrix. Its *transposed matrix* A^T is obtained by interchanging the rows and columns of A, that is, $A^T = \{a_{ji}\} \in \mathbb{R}^{m\times n}$.

The product of two matrices $A \in \mathbb{R}^{n\times p}$ and $B \in \mathbb{R}^{p\times m}$ is denoted as $C = AB \in \mathbb{R}^{n\times m}$, that is,

$$
\begin{aligned}
C = AB &= \begin{bmatrix} a_{11} & a_{12} & \ldots & a_{1p} \\ a_{21} & a_{22} & \ldots & a_{2p} \\ \vdots & \vdots & \ddots & \vdots \\ a_{n1} & a_{n2} & \ldots & a_{np} \end{bmatrix} \cdot \begin{bmatrix} b_{11} & b_{12} & \ldots & b_{1m} \\ b_{21} & b_{22} & \ldots & b_{2m} \\ \vdots & \vdots & \ddots & \vdots \\ b_{p1} & b_{p2} & \ldots & b_{pm} \end{bmatrix} \\
&= \begin{bmatrix} \sum_{k=1}^{p} a_{1k}b_{k1} & \sum_{k=1}^{p} a_{1k}b_{k2} & \ldots & \sum_{k=1}^{p} a_{1k}b_{km} \\ \sum_{k=1}^{p} a_{2k}b_{k1} & \sum_{k=1}^{p} a_{2k}b_{k2} & \ldots & \sum_{k=1}^{p} a_{2k}b_{km} \\ \vdots & \vdots & \ddots & \vdots \\ \sum_{k=1}^{p} a_{nk}b_{k1} & \sum_{k=1}^{p} a_{nk}b_{k2} & \ldots & \sum_{k=1}^{p} a_{nk}b_{km} \end{bmatrix}.
\end{aligned} \tag{A.6}
$$

As in vectors, the product of matrices satisfies the following properties:

- $(AB)^T = B^T A^T$, for all $A \in \mathbb{R}^{n\times p}$ and $B \in \mathbb{R}^{p\times m}$,
- $AB \neq BA$, in general,
- $A(B + C) = AB + AC$, for all $A \in \mathbb{R}^{n\times p}$ and $B, C \in \mathbb{R}^{p\times m}$,
- $ABC = A(BC) = (AB)C$, for all $A \in \mathbb{R}^{n\times p}$, $B \in \mathbb{R}^{p\times m}$, and $C \in \mathbb{R}^{m\times r}$.

A matrix $A \in \mathbb{R}^{n\times m}$ is a *square matrix* if $n = m$. Thus, a square matrix $A \in \mathbb{R}^{n\times n}$ is called a *symmetric matrix* if it is equal to its transpose, that is, $A = A^T$, while it is called an *antisymmetric matrix* if $A = -A^T$. A property of an antisymmetric matrix is

$$x^T Ax = 0, \quad \text{for all } x \in \mathbb{R}^n. \tag{A.7}$$

A square matrix $A = \{a_{ij}\} \in \mathbb{R}^{n\times n}$ is called a *diagonal matrix* if $a_{ij} = 0$ for all $i \neq j$, and it is denoted $\text{diag}\{a_{11}, a_{22}, \ldots, a_{nn}\} \in \mathbb{R}^{n\times n}$, for example,

$$\text{diag}\{a_{11}, a_{22}, \ldots, a_{nn}\} = \begin{bmatrix} a_{11} & 0 & \ldots & 0 \\ 0 & a_{22} & \ldots & 0 \\ \vdots & \vdots & \ddots & \vdots \\ 0 & 0 & \ldots & a_{nm} \end{bmatrix} \in \mathbb{R}^{n\times n}. \tag{A.8}$$

Therefore, any diagonal matrix is also a symmetrical matrix. For the case where $a_{11} = a_{22} = \ldots = a_{nn} = \alpha$, the diagonal matrix will be denoted $\text{diag}\{\alpha\} \in \mathbb{R}^{n\times n}$, where $\alpha \in \mathbb{R}$. Two useful diagonal matrices are the *identity matrix* of dimension n, defined as $I = \text{diag}\{1\} \in \mathbb{R}^{n\times n}$, and the *null matrix* of dimension n, defined as $0 = \text{diag}\{0\} \in \mathbb{R}^{n\times n}$.

Any matrix A satisfies

$$A^T A = AA^T = I. \tag{A.9}$$

The *trace* of a square matrix $A \in \mathbb{R}^{n\times n}$, represented as $\mathrm{tr}(A)$ or $\mathrm{trace}(A)$, is the sum of the elements of the main diagonal, that is,

$$\mathrm{tr}(A) = \sum_{i=1}^{n} a_{ii}. \tag{A.10}$$

A square matrix $A \in \mathbb{R}^{n\times n}$ is called a *singular matrix* if its determinant is zero, that is, if $\det[A] = 0$, otherwise it is called a *nonsingular matrix.* A feature of a singular array is that it has no inverse.

Theorem 1 (Sylvester theorem). *Let P be a square matrix of dimension n. Define*

$$A = \{a_{ij}\} = \frac{P + P^T}{2}. \tag{A.11}$$

The matrix P will be positive definite if and only if

$$\det[a_{11}] > 0, \det\begin{bmatrix} a_{11} & a_{12} \\ a_{21} & a_{22} \end{bmatrix} > 0, \ldots, \det[A] > 0. \tag{A.12}$$

A square matrix $A \in \mathbb{R}^{n\times n}$, without being symmetric, is also positive definite if

$$x^T A x > 0, \qquad \forall\, x \in \mathbb{R}^n\,,\ x \neq 0 \in \mathbb{R}^n, \tag{A.13}$$

where $A > 0$ means that the matrix A is positive definite and not that A is greater than zero. Likewise, any symmetric and positive definite matrix $A = A^T > 0$ is nonsingular, so therefore, its inverse exists. The sum of two positive definite matrices also results in a positive definite matrix, while the product of two positive definite symmetric matrices $A = A^T > 0$ and $B = B^T > 0$ is not generally symmetric and also not positive definite; however, the resulting matrix is not singular.

A square matrix $A \in \mathbb{R}^{n\times n}$ that is not necessarily symmetric is *positive semidefinite* if

$$x^T A x \geq 0, \qquad \forall\, x \in \mathbb{R}^n. \tag{A.14}$$

Then it can be said that a square matrix $A \in \mathbb{R}^{n\times n}$ is *negative definite* if $-A$ is positive definite, while it is *negative semidefinite* if $-A$ is positive semidefinite.

For every square matrix $A \in \mathbb{R}^{n\times n}$ there are n *eigenvalues* denoted by $\lambda_1\{A\}$, $\lambda_2\{A\}, \ldots, \lambda_n\{A\}$. The value properties of the matrix A satisfy

$$\det[\lambda_i\{A\}I - A] = 0, \qquad \text{for } i = 1, 2, \ldots, n, \tag{A.15}$$

where $I \in \mathbb{R}^{n\times n}$ is an identity matrix of dimension n. For the case of a symmetric matrix $A = A^T \in \mathbb{R}^{n\times n}$, for their eigenvalues we have:

- $\lambda_1\{A\}, \lambda_2\{A\}, \ldots, \lambda_n\{A\}$ are real numbers,

- denoting the maximum and minimum eigenvalues of A as $\lambda_{max}\{A\}$ and $\lambda_{min}\{A\}$, respectively, the *Rayleigh–Ritz theorem* states that for all $x \in \mathbb{R}^n$ we have

$$\lambda_{max}\{A\}\|x\|^2 \geq x^T Ax \geq \lambda_{min}\{A\}\|x\|^2. \tag{A.16}$$

A square matrix $A \in \mathbb{R}^{n\times n}$ is positive definite if and only if the eigenvalues of $A + A^T$ are positive, that is, $\lambda_i\{A + A^T\} > 0$ for $i = 1, 2, \ldots, n$. Furthermore, a symmetric matrix $A = A^T \in \mathbb{R}^{n\times n}$ is positive definite if and only if $\lambda_i\{A\} > 0$ for $i = 1, 2, \ldots, n$.

Given the n eigenvalues of a square matrix $A \in \mathbb{R}^{n\times n}$, the trace and determinant of the matrix A can be calculated as

$$\text{tr}(A) = \sum_{i=1}^{n} \lambda_i\{A\}, \tag{A.17}$$

$$\det(A) = \prod_{i=1}^{n} \lambda_i\{A\}. \tag{A.18}$$

The *range* of an array A, denoted as rank(A), is the order of the largest to the smallest distinct from zero. Therefore, the range cannot be greater than the number of rows or columns. Another way to calculate the rank of an array A is by obtaining the number of lines (rows or columns of the said matrix) which are linearly independent.

A.5 Spectral norm

The *spectral norm* of a matrix $A \in \mathbb{R}^{n\times m}$, represented as $\|A\|$, is defined as follows:

$$\|A\| = \sqrt{\lambda_{max}\{A^T A\}}, \tag{A.19}$$

where $\lambda_{max}\{A^T A\}$ denotes the maximum eigenvalue of the symmetric matrix $A^T A \in \mathbb{R}^{m\times m}$. For the particular case of symmetric matrices $A = A^T \in \mathbb{R}^{n\times n}$, we have:

- $\|A\| = \max_i |\lambda_i\{A\}|$,
- $\|A^{-1}\| = \frac{1}{\min_i |\lambda_i\{A\}|}$.

For the case in which the matrix A is symmetric and positive definite, the absolute value symbol of the previous expressions is redundant.

The spectral norm satisfies the following properties:

- $\|A\| = 0$, if and only if $A = 0 \in \mathbb{R}^{n\times m}$,
- $\|A\| > 0$, for all $A \in \mathbb{R}^{n\times m}$ con $A \neq 0 \in \mathbb{R}^{n\times m}$,
- $\|A + B\| \leq \|A\| + \|B\|$, for all $A, B \in \mathbb{R}^{n\times m}$,
- $\|\alpha A\| = |\alpha|\ \|A\|$, for all $\alpha \in \mathbb{R}$ and $A \in \mathbb{R}^{n\times m}$,
- $\|A^T B\| \leq \|A\|\|B\|$, for all $A, B \in \mathbb{R}^{n\times m}$.

Other useful inequalities are the following. Consider the matrix $A \in \mathbb{R}^{n \times m}$ and the vector $x \in \mathbb{R}^m$. The Euclidean norm of the vector Ax satisfies

$$\|Ax\| \leq \|A\| \, \|x\|, \tag{A.20}$$

where logically $\|A\|$ is the spectral norm of A and $\|x\|$ is the Euclidean norm of x. Likewise, let $y \in \mathbb{R}^n$. The absolute value of $y^T Ax$ satisfies

$$\|y^T Ax\| \leq \|A\| \, \|y\| \, \|x\|. \tag{A.21}$$

A.6 P-norms

Now consider the class of *p-norms* defined as

$$\|x\|_p = \left(|x_1|^p + \ldots + |x_n|^p\right)^{1/p}, \qquad \forall\, x \in \mathbb{R}^n,\ 1 \leq p < \infty, \tag{A.22}$$

and

$$\|x\|_\infty = \max_i |x_i|. \tag{A.23}$$

The three most common rules are $\|x\|_1$, $\|x\|_\infty$, and the Euclidean norm $\|x\|_2$. All p-norms are equivalent, in the sense that if $\|\cdot\|_\alpha$ and $\|\cdot\|_\beta$ are two different p-norms, then there exist positive constants c_1 and c_2 such that

$$c_1\|x\|_\alpha \leq \|x\|_\beta \leq c_2\|x\|_\alpha, \qquad \forall\, x \in \mathbb{R}^n. \tag{A.24}$$

For the particular case of the three most common standards, inequality (A.24) becomes

$$\|x\|_2 \leq \|x\|_1 \leq \sqrt{n}\|x\|_2, \tag{A.25}$$

$$\|x\|_\infty \leq \|x\|_2 \leq \sqrt{n}\|x\|_\infty, \tag{A.26}$$

$$\|x\|_\infty \leq \|x\|_1 \leq n\|x\|_\infty, \tag{A.27}$$

where $x \in \mathbb{R}^n$.

Two important results obtained from p-norms are the so-called *Hölder inequality*,

$$|x^T y| \leq \|x\|_p \, \|y\|_q, \qquad \text{with } \frac{1}{p} + \frac{1}{q} = 1, \qquad \forall\, x, y \in \mathbb{R}^n, \tag{A.28}$$

and the so-called *Minkowski inequality*

$$\|x + y\|_p \leq \|x\|_p + \|y\|_p, \qquad \text{with } p \geq 1, \qquad \forall\, x, y \in \mathbb{R}^n. \tag{A.29}$$

For the case of a matrix $A \in \mathbb{R}^{n \times m}$ with real definite elements and a vector $x \in \mathbb{R}^m$, the matrix defines a linear mapping $y = Ax$, which goes from $\mathbb{R}^m$ to $\mathbb{R}^n$. The induced p-norm of $A \in \mathbb{R}^{n\times m}$ is given by

$$\|A\|_p = \sup_{x\neq 0} \frac{\|Ax\|_p}{\|x\|_p} = \max_{\|x\|_p=1} \|Ax\|_p, \tag{A.30}$$

so in the case of $p = 1, 2, \infty$, Eq. (A.30) results in

$$\|A\|_1 = \max_j \sum_{i=1}^{n} |a_{ij}|, \tag{A.31}$$

$$\|A\|_2 = \left[\lambda_{max}\{A^T A\}\right]^{1/2}, \tag{A.32}$$

$$\|A\|_\infty = \max_i \sum_{j=1}^{m} |a_{ij}|. \tag{A.33}$$

Some more common properties for real arrays A and B of dimensions $n \times m$ and $m \times l$, respectively, are

$$\frac{1}{\sqrt{m}}\|A\|_\infty \leq \|A\|_2 \leq \sqrt{n}\|A\|_\infty, \tag{A.34}$$

$$\frac{1}{\sqrt{n}}\|A\|_1 \leq \|A\|_2 \leq \sqrt{m}\|A\|_1, \tag{A.35}$$

$$\|A\|_2 \leq \sqrt{\|A\|_1 + \|A\|_\infty}, \tag{A.36}$$

$$\|AB\|_p \leq \|A\|_p \, \|B\|_p. \tag{A.37}$$

A.7 Dyadic product, cross product, and antisymmetric matrix

The *dyadic* is a cross product defined as

$$x \circ y = xy^T = \begin{bmatrix} x_1 \\ x_2 \\ \vdots \\ x_n \end{bmatrix} \begin{bmatrix} y_1 & y_2 & \cdots & y_n \end{bmatrix} \in \mathbb{R}^{n\times n} \qquad \forall x, y \in \mathbb{R}^n. \tag{A.38}$$

Defining $x, y, z \in \mathbb{R}^n$, it can be verified that

$$(x \circ y)z = (y^T z)x, \tag{A.39}$$

$$(x \circ y)^T z = (x^T z)y. \tag{A.40}$$

Likewise, $x \circ x = xx^T$, which forms a symmetric positive semidefinite matrix.

For the case of vectors in $\mathbb{R}^3$, there is also another product of vectors, called *cross product*, defined as follows. Let $x, y, z \in \mathbb{R}^3$. Then

$$z = x \times y = \begin{bmatrix} x_1 \\ x_2 \\ x_3 \end{bmatrix} \times \begin{bmatrix} y_1 \\ y_2 \\ y_3 \end{bmatrix} = \begin{bmatrix} x_2 y_3 - x_3 y_2 \\ x_3 y_1 - x_1 y_3 \\ x_1 y_2 - x_2 y_1 \end{bmatrix}. \tag{A.41}$$

Another way to express the cross product is through the use of an antisymmetric matrix, which is defined as follows. Let $x \in \mathbb{R}^3$. The *antisymmetric matrix* $\hat{x}$ is composed as

$$\hat{x} = \begin{bmatrix} 0 & -x_3 & x_2 \\ x_3 & 0 & -x_1 \\ -x_2 & x_1 & 0 \end{bmatrix}. \tag{A.42}$$

Using (A.42), the cross product can be rewritten as

$$x \times y = \hat{x} y. \tag{A.43}$$

Using the three types of products with vectors (point, vector, and dyadic), given the vectors $x, y \in \mathbb{R}^3$, we have

$$x \times (x \times y) = (x \circ x)y - (x^T x)y, \tag{A.44}$$

which is equivalent to

$$\hat{x}\hat{x}y = xx^T y - x^T xy, \tag{A.45}$$

so it can also be concluded that

$$\hat{x}^2 = \hat{x}\hat{x} = xx^T - \|x\|^2 I, \tag{A.46}$$

where $I \in \mathbb{R}^{3\times 3}$ is the identity matrix.

Going back to the antisymmetric matrix, given the vectors $x, y, z \in \mathbb{R}^3$, the antisymmetric matrix has the following properties:

$$\hat{x}^T = -\hat{x}, \tag{A.47}$$

$$\hat{x}(y + z) = \hat{x}y + \hat{x}z, \tag{A.48}$$

$$\hat{x}y = -\hat{y}x, \tag{A.49}$$

$$\hat{x}x = 0 \in \mathbb{R}^3, \tag{A.50}$$

$$y^T \hat{x} y = 0, \tag{A.51}$$

$$x^T \hat{y} z = z^T \hat{x} y, \tag{A.52}$$

$$\hat{x}\hat{y} = yx^T - x^T y I, \tag{A.53}$$

$$\widehat{(\hat{x}y)} = yx^T - xy^T. \tag{A.54}$$

A.8 Topological concepts

A.8.1 Sets

A *set* is a collection of objects, points, or items. Thus, if A is a set and x is an element of the set A, this relation is written as $x \in A$. If A and B are two sets and each element of the set A is also an element of the set B, then we say that the set A is included in set B. Therefore, the set A can be viewed as a *subset* of B, which is written mathematically as $A \subset B$ or $B \supset A$. If a set has no elements, it is an empty set, $\emptyset$. Therefore, each set with elements contains an empty set. The *union* and *intersection* of two given sets A and B are defined respectively as

$$A \cup B = \{x \mid x \in A \quad \text{or} \quad x \in B\}, \tag{A.55}$$

$$A \cap B = \{x \mid x \in A \quad \text{and} \quad x \in B\}. \tag{A.56}$$

Consider two sets A and B. The *difference* $A \setminus B$ is the set of all points in A that do not belong to B. Consider now two nonempty sets A and B. Then the *Cartesian product* $A \times B$ of the sets A and B is a set of all ordered pairs of the form (a, b) with $a \in A$ and $b \in B$

$$A \times B = \{(a, b) \mid a \in A \quad \text{and} \quad b \in B\}. \tag{A.57}$$

A subset $S \subset \mathbb{R}^n$ is called an *open set* if for every vector $x \in S$ one can find an ϵ neighborhood of x

$$N(x, \epsilon) = \{z \in \mathbb{R}^n \mid \|z - x\| < \epsilon\} \tag{A.58}$$

such that $N(x, \epsilon) \subset S$. A set S is called a *closed set* if and only if each convergent sequence $\{x_k\}$ with elements in S converges to a point in S. A set S is called a *bounded set* if there exists an $r > 0$ such that

$$\|x\| \leq r, \qquad \forall x \in S. \tag{A.59}$$

A set S is called a *compact set* if it is closed and bounded.

A point p is a *boundary point* of a set S if every neighborhood of p contains at least one point in S and one point that does not belong to S. The set of all boundary points of S, denoted ∂S, is called the *boundary of* S. A closed set contains all its boundary points, whereas an open set does not contain its boundary points.

The *inside* of a set S is given by $S - \partial S$, so an open set is equal to its interior. The *lock* of a set S, denoted by $\bar{S}$, is the union of S and its border. Thus, a closed set is equal to its lock. An open set S is *connected* if every pair of a fictitious line can join points of S in S.

A set S is called a *region* if it is the union of an open set connected with some, none, or all of its boundary points. If no boundary point is included, the region is called an *open region* or *domain*.

A set S is a *convex set* if for every $x, y \in S$ and every real number θ (where $0 < \theta < 1$) the point

$$\theta x + (1 - \theta)y \in S. \tag{A.60}$$

If $x \in X \subset \mathbb{R}^n$ and $y \in Y \subset \mathbb{R}^m$, it can be said that the pair (x, y) belongs to the product set

$$X \times Y \subset \mathbb{R}^n \times \mathbb{R}^m. \tag{A.61}$$

Some of the above concepts will now be extrapolated to $\mathbb{R}^n$. Consider the set $A \subset \mathbb{R}^n$. The *neighborhood* of a point $p \in A \subset \mathbb{R}^n$ is given by the set $B_r(p)$ defined as

$$B_r(p) = \{x \in \mathbb{R}^n \mid \|x - p\| < r\}, \tag{A.62}$$

which is often called an *open ball* with center at p and radius r. The set $A \subset \mathbb{R}^n$ is said to be an *open set* if for any $p \in A$ one can find a neighborhood $B_r(p) \subset A$.

The set $A \subset \mathbb{R}^n$ is said to be a *convex set* if for any $x_1, x_2 \in A$

$$\theta x_1 + (1 - \theta)x_2 \in A, \qquad 0 \leq \theta \leq 1. \tag{A.63}$$

Let A be a given set. Then a *structure topology* or *topology* in A is a collection of subsets of A called open sets that satisfy the following conditions:

- The union of any number of open sets results in an open set.
- The intersection of any finite number of open sets is an open set.
- The sets A and empty $\emptyset$ are open.

Thus, a set A with a topology is called a *topological space*.

A.8.2 Metric spaces

In analyzing real or complex numbers, many results depend only on the idea of the distance between two numbers x and y. Metric spaces form a natural generalization of distance.

A metric space is a pair (X, d) of a nonempty set X and a distance metric or function $d : X \times X \to \mathbb{R}$ such that for all $x, y, z \in X$, the following conditions hold:

(i) $d(x, y) = 0$ if and only if $x = y$,
(ii) $d(x, y) = d(y, x)$,
(iii) $d(x, z) \leq d(x, y) + d(y, z)$.

Property (iii) is the previously seen triangle inequality. Setting $x = z$ in (iii) and taking into account (i) and (ii), we have $d(x, x) = 0 \leq d(x, y) + d(y, x) = 2d(x, y)$, from which we see that $d(x, y) \geq 0 \; \forall x, y \in X$.

A.8.3 Linear independence

It is said that a finite set of vectors $x_i \in \mathbb{R}^n$ is linearly dependent if there exists a set corresponding to scalar values $\alpha_i \in \mathbb{R}$ (with $i = 1, 2, ..., n$), not all zero, such that

$$\sum_{i=1}^{n} \alpha_i x_i = 0. \tag{A.64}$$

Otherwise, the set of vectors x_i is said to be linearly independent.

A.8.4 Sequence convergence

A *sequence of vectors* $x_0, x_1, \ldots, x_k, \ldots$ in $\mathbb{R}^n$, denoted by $\{x_k\}$, is said to converge to a limit vector if

$$\|x_k - x\| \to 0 \qquad \text{when } k \to \infty, \tag{A.65}$$

which is equivalent to saying that, given any $\epsilon > 0$, there is a value N such that

$$\|x_k - x\| \to \epsilon \qquad \forall\, k \geq N, \tag{A.66}$$

which implies that $d(x_n, x) < \epsilon$. Therefore, it can be written that $x = \lim x_n$, or $x_n \to x$.

A vector x is an *accumulation point* of a sequence $\{x_k\}$ if there is a subsequence of $\{x_k\}$ that converges to x, that is, if there exists an infinite subset K of the nonnegative integers such that $\{x_k\}_{k \in K}$ converges to x. A *bounded sequence* $\{x_k\} \in \mathbb{R}^n$ has at least one accumulation point at $\mathbb{R}^n$.

A sequence of real numbers $\{r_k\}$ is called an *increasing sequence* (*monotonically increasing* or not decreasing if $r_k \leq r_{k+1}$ $\forall\, k$ and *strictly increasing* if $r_k < r_{k+1}$). The *decreasing sequence* (*monotonically decreasing* or not increasing if $r_k \geq r_{k+1}$ and *strictly decreasing* if $r_k > r_{k+1}$) is similarly defined.

An increasing sequence of real numbers that is bounded superiorly converges to a real number. Similarly, a sequence of decreasing real numbers that is bounded from below converges to a real number.

A sequence x_n in a metric space (X, d) is called a *Cauchy sequence* if for every real number $\chi > 0$ there exists an integer N such that $d(x_n, x_m) < \chi$ with the condition that $n, m \geq N$. Thus, it is easy to see that each convergent sequence is a Cauchy sequence, but the inverse is not necessarily true.

A.9 Functions

A.9.1 Continuous functions

A function f that *maps* one set S_1 to another set S_2 is represented as $f : S_1 \to S_2$. A function $f : \mathbb{R}^n \to \mathbb{R}^m$ is said to be *continuous* at a point x if $f(x_k) \to f(x)$ as long as $x_k \to x$. Equivalently, f is continuous at x if given $\epsilon > 0$ there exists a $\delta > 0$ such that

$$\|x - y\| < \delta \quad \Rightarrow \quad \|f(x) - f(y)\| < \epsilon. \tag{A.67}$$

A function f is continuous on a set S if it is continuous for each point of S, and it is *uniformly continuous* on S if given $\epsilon > 0$, there exists a $\delta > 0$ (depending only on ϵ) such that inequality (A.67) holds for all $x, y \in S$.

It should be noted that "uniform continuity" is defined for a set, while "continuity" is defined as a point. For uniform continuity, the same constant δ works for all points within the set. If f is uniformly continuous on a set S, this continues in S. The opposite of the above statement is not always true. However, if S is a compact set, then continuity and uniform continuity over S are equivalent.

A function f is called a *soft function* (or of class C^∞) if it can be infinitely differentiable and its derivatives are continuous.

A function f is said to be an *injective function* if for every $y \in Y$ there exists at most one $x \in X$ such that $y = f(x)$. On the other hand, a function f is called a *surjective function* if for every $y \in Y$ there is at least one $x \in X$ such that $y = f(x)$. So, the function f is called a *bijective function* if it is injective and surjective, that is, one by one, where a function f defined on a set S is said to be *one by one* in S if any value of $x, y \in S$ with $x \neq y$ such that

$$f(x) \neq f(y). \tag{A.68}$$

The function

$$(a_1 f_1 + a_2 f_2)(\cdot) = a_1 f_1(\cdot) + a_2 f_2(\cdot) \tag{A.69}$$

is continuous for any scalar value of a_1 and a_2 and any continuous functions $f_1(\cdot)$ and $f_2(\cdot)$. If S_1, S_2, and S_3 are three given sets and $f_1 : S_1 \to S_2$ and $f_2 : S_2 \to S_3$ are functions, then the function $f_2 \circ f_1 : S_1 \to S_3$, defined as

$$(f_2 \circ f_1)(\cdot) = f_2(f_1(\cdot)), \tag{A.70}$$

is called the *composition* of f_1 and f_2. The composition of two continuous functions is also continuous.

If $S \subset \mathbb{R}^n$ and $f : S \to \mathbb{R}^m$, then the set of $f(x)$ such that $x \in S$ is called the *image of S under f* and is represented by $f(S)$.

If f is a continuous function defined on the compact set S, then $f(S)$ is also compact. So a continuous function defined on a compact set is bounded. Also, if f is a scalar function, that is, $f : S \to \mathbb{R}$, then there exist points p and q in the compact set S such that

$$f(x) \leq f(p), \tag{A.71}$$

$$f(x) \geq f(q) \tag{A.72}$$

for all $x \in S$. If f is a continuous function, defined on a connected set S, then $f(S)$ is also connected. If $f : S \to \mathbb{R}^m$ is a continuous function, one to one in a compact set $S \subset \mathbb{R}^n$, then f has a continuous inverse f^{-1} on $f(S)$. The composition of f and f^{-1} is the identity, that is, $f^{-1}(f(x)) = x$.

A function $f : \mathbb{R} \to \mathbb{R}^n$ is said to be *piecewise continuous* on an interval $J \subset \mathbb{R}$ if for each bounded subinterval $J_0 \subset J$, f is continuous for all $x \in J_0$, except possibly at a finite number of points where f can have discontinuities. Furthermore, at each point of discontinuity x_0, the limit of the right-hand side

$$\lim_{h\to 0} f(x_0 + h) \tag{A.73}$$

and the limit of the left-hand side

$$\lim_{h\to 0} f(x_0 - h) \tag{A.74}$$

exists; that is, the function has a finite jump at x_0. Two topological spaces X and Y are *homeomorphic* if there is a function f that does a one-to-one mapping between the two spaces, and vice versa. So a function $f : X \to Y$ is a *homeomorphism* if and only if:

- f is bijective,
- f is continuous,
- f^{-1} is continuous.

A function $f : X \to Y$ is said to be a *linear operator* (or a *linear mapping* or a *linear transformation*) if and only if for any value of $x_1, x_2 \in X$ and any value of $\lambda, \mu \in \mathbb{R}$

$$f(\lambda x_1 + \mu x_2) = \lambda f(x_1) + \mu f(x_2). \tag{A.75}$$

Also, the function f is said to be a *bounded linear operator* if there exists a constant M such that

$$\|f(x)\| \le M\|x\| \qquad \forall x \in X. \tag{A.76}$$

A.9.2 Differentiable functions

A function $f : \mathbb{R} \to \mathbb{R}$ is said to be a *differentiable function* on x if the limit

$$f'(x) = \lim_{h\to 0} \frac{f(x+h) - f(x)}{h} \tag{A.77}$$

exists. The limit $f'(x)$ is called *the derivative of f in x*. If the derivative exists, then

$$f(x+h) - f(x) = f'(x)h + r(h), \tag{A.78}$$

where the remainder $r(h)$ has a small value, such that

$$\lim_{h\to 0} \frac{r(h)}{h} = 0. \tag{A.79}$$

A function $f : \mathbb{R}^n \to \mathbb{R}^m$ is differentiable at a point x if f is defined on an open set $D \subset \mathbb{R}^n$ containing x and the limit

$$\lim_{h\to 0} \frac{||f(x+h) - f(x) - f'(x)h||}{||h||} = 0 \tag{A.80}$$

exists. Note that $h \in \mathbb{R}^n$. If $||h||$ is small enough, then $x + h \in D$, since D is an open set and $f(x+h) \in \mathbb{R}^m$.

A function $f : \mathbb{R}^n \to \mathbb{R}^m$ is said to be *continuously differentiable* at a point x_0 if the partial derivatives $\frac{\partial f_i}{\partial x_j}$ exist and are continuous at x_0 for $1 \le i \le m$ and $1 \le j \le n$. A function f is continuously differentiable on a set S if it is continuously differentiable for each point of S.

For a continuously differentiable function $f : \mathbb{R}^n \to \mathbb{R}$, the row vector of $\frac{\partial f}{\partial x}$ is defined as

$$\frac{\partial f}{\partial x} = \begin{bmatrix} \frac{\partial f}{\partial x_1} & \cdots & \frac{\partial f}{\partial x_n} \end{bmatrix}. \tag{A.81}$$

The *gradient vector*, represented by $\nabla f(x)$, is

$$\nabla f(x) = \left[\frac{\partial f}{\partial x}\right]^T. \tag{A.82}$$

For a continuously differentiable function $f : \mathbb{R}^n \to \mathbb{R}^m$, the *Jacobian matrix* $[\frac{\partial f}{\partial x}]$ is given by a matrix of dimension $m \times n$, where the element of the i-th row and the j-th column is given by

$$\frac{\partial f_i}{\partial x_j}. \tag{A.83}$$

Suppose $S \subset \mathbb{R}^n$ is open, f maps S on $\mathbb{R}^m$, f is continuously differentiable on $x_0 \in S$, g maps an open set given by $f(S)$ into $\mathbb{R}^k$, and g is continuously differentiable on $f(x_0)$. Then, mapping h from S to $\mathbb{R}^k$, defined by $h(x) = g(f(x))$, it is continuously differentiable at x_0 and its Jacobian matrix is given by the *string rule*:

$$\left.\frac{\partial h}{\partial x}\right|_{x=x_0} = \left.\frac{\partial g}{\partial f}\right|_{f=f(x_0)} \cdot \left.\frac{\partial f}{\partial x}\right|_{x=x_0}. \tag{A.84}$$

A.9.3 Mean value theorem

If x, y are two distinct points on $\mathbb{R}^n$, then the line segment $L(x, y)$ that joins point x and point y is

$$L(x, y) = \{z \mid z = \theta x + (1-\theta)y,\ 0 < \theta < 1\}. \tag{A.85}$$

Theorem 2. *Consider the function $f : \mathbb{R}^n \to \mathbb{R}$, which is continuously differentiable for each point x of the open set $S \subset \mathbb{R}^n$. Let x, y be two points of S such that the line segment $L(x, y) \subset S$. Then there exists a point z of $L(x, y)$ such that*

$$f(y) - f(x) = \frac{\partial f}{\partial x}|_{x=z}(y - x). \tag{A.86}$$

A.9.4 Implicit function theorem

Consider the function $f : \mathbb{R}^n \times \mathbb{R}^m \to \mathbb{R}^n$, which is continuously differentiable for each point (x, y) of an open set $S \subset \mathbb{R}^n \times \mathbb{R}^m$. Let (x_0, y_0) be a point in S for which $f(x_0, y_0) = 0$ and for which the matrix Jacobian $[\frac{\partial f}{\partial x}](x_0, y_0)$ is not singular. Then, there are neighborhoods $U \subset \mathbb{R}^n$ of x_0 and $V \subset \mathbb{R}^m$ of y_0 such that for each $y \in V$, the equation $f(x, y) = 0$ has a unique solution at $x \in US$. Furthermore, this solution can be given as $x = g(y)$, where g is continuously differentiable at $y = y_0$.

A.9.5 Gronwall–Bellman inequality

Lemma. *Let $\lambda : [a, b] \to \mathbb{R}$ be a continuous function and let $\mu : [a, b] \to \mathbb{R}$ a non-continuous negative function. If a continuous function $y : [a, b] \to \mathbb{R}$ satisfies*

$$y(t) \le \lambda(t) + \int_a^t \mu(s)y(s)ds \tag{A.87}$$

for $a \le t \le b$, then in the same interval

$$y(t) \le \lambda(t) + \int_a^t \lambda(s)\mu(s)e^{\int_s^t \mu(\tau)d\tau}ds. \tag{A.88}$$

For the case where $\lambda(t) \equiv \lambda$ is a constant, we have

$$y(t) \le \lambda e^{\int_a^t \mu(\tau)d\tau}. \tag{A.89}$$

If moreover $\mu(t) \equiv \mu \ge 0$ is a constant, then

$$y(t) \le \lambda e^{\mu(t-a)}. \tag{A.90}$$

Proof. Let $z(t) = \int_a^t \mu(s)y(s)ds$ and $v(t) = z(t) + \lambda(t) - y(t) \ge 0$. Then z is differentiable and

$$\dot{z} = \mu(t)y(t) = \mu(t)z(t) + \mu(t)\lambda(t) - \mu(t)v(t). \tag{A.91}$$

This is a linear and scalar equation of state with the following transition state function:

$$\phi(t, s) = e^{\int_s^t \mu(\tau)d\tau}. \tag{A.92}$$

Since $z(a) = 0$, we have

$$z(t) = \int_a^t \phi(t,s)[\mu(s)\lambda(s) - \mu(s)v(s)]ds. \tag{A.93}$$

The term

$$\int_a^t \phi(t,s)\mu(s)v(s)ds \tag{A.94}$$

is nonnegative. Therefore,

$$z(t) \leq \int_a^t e^{\int_s^t \mu(\tau)d\tau}\mu(s)\lambda(s)ds. \tag{A.95}$$

Since $y(t) \leq \lambda(t) + z(t)$, this completes the proof for the general case. In the particular case of $\lambda(t) \equiv \lambda$, we have

$$\int_a^t \mu(s)e^{\int_s^t \mu(\tau)d\tau}ds = -\int_a^t \frac{d}{ds}\left[e^{\int_s^t \mu(\tau)d\tau}\right]ds \tag{A.96}$$

$$= -\left[e^{\int_s^t \mu(\tau)d\tau}\right]\Big|_{s=a}^{s=t} \tag{A.97}$$

$$= -1 + e^{\int_a^t \mu(\tau)d\tau}, \tag{A.98}$$

which proves the lemma when λ is a constant. We get the proof when λ and μ are constants integrating (A.95).

A.10 Contraction mapping

Consider an equation of the form

$$y = T(x) \qquad \text{for } x \in \mathbb{R}^n, \tag{A.99}$$

where the solution x^* of said equation is called a *fixed point* of the mapping T since T leaves x^* invariant, that is, $T(x^*) = x^*$.

A classic way to find a fixed point is by using the method of successive approximations. It starts with an initial vector test x_1 and we calculate $x_2 = T(x_1)$. Continuing in this way iteratively, successive vectors are computed as $x_{k+1} = T(x_k)$. Finally, we check if the sequence $\{x_k\}$ converges to x^*.

A *linear vector space* χ over the field $\mathbb{R}^n$ is a set of elements $x, y, \ldots$ called vectors, such that for any vector $x, y, z \in \chi$

$$x + y \in \chi, \tag{A.100}$$

$$x + y = y + x, \tag{A.101}$$

$$(x + y) + z = x + (y + z) \tag{A.102}$$

and there exists a zero vector $0 \in \chi$ such that $x + 0 = x$ for all $x \in \chi$. Also, for any number $\alpha, \beta \in \mathbb{R}$ and all $x, y \in \chi$,

$$\alpha x \in \chi, \tag{A.103}$$
$$1 \cdot x = x, \tag{A.104}$$
$$0 \cdot x = 0, \tag{A.105}$$
$$(\alpha\, \beta)x = \alpha(\beta x), \tag{A.106}$$
$$\alpha(x + y) = \alpha x + \alpha y, \tag{A.107}$$
$$(\alpha + \beta)x = \alpha x + \beta x. \tag{A.108}$$

A linear space χ is called a *normed linear space* if for each vector $x \in \chi$, there exists a scalar function called *norm* $|x|$ that satisfies:

- $|x| \geq 0$ for all $x \in \chi$,
- $|x| = 0$ if and only if $x = 0$,
- $|x + y| \leq |x| + |y|$ for all $x, y \in \chi$,
- $|\alpha x| = |\alpha||x|$ for all $\alpha \in \mathbb{R}$ and $x \in \chi$.

Hereafter, if it is not clear in the context when $\|\cdot\|$ is a norm on χ or a norm on $\mathbb{R}^n$, then we will write $\|\cdot\|_\chi$ for a norm in χ.

A sequence $\{x_k\} \in \chi$ in a normed linear space *converges* to $x \in \chi$ if

$$\|x_k - x\| \to 0 \qquad \text{when} \qquad k \to \infty. \tag{A.109}$$

A set $S \subset \chi$ is a *closed set* if and only if every convergent sequence with elements in S has its limit in S.

A sequence $\{x_k\} \in \chi$ is called a *Cauchy sequence* if

$$\|x_k - x_m\| \to 0 \qquad \text{when} \qquad k, m \to \infty. \tag{A.110}$$

Every convergent sequence is Cauchy, but not every Cauchy sequence is convergent.

A normed linear space χ is *complete* if each Cauchy sequence in χ converges to a vector in χ. A complete normed linear space is called *Banach space*.

Contraction mapping theorem: Let S be a closed subset of a Banach space χ and let T be a mapping from S to S. Suppose that

$$\|T(x) - T(y)\| \leq \rho\|x - y\|, \qquad \forall\, x, y \in S,\ 0 \leq \rho < 1. \tag{A.111}$$

Then:

a) there is a unique vector $x^* \in S$ that satisfies $x^* = T(x^*)$,
b) x^* can be obtained by the method of successive approximations, starting from any arbitrary initial vector in S.

Proof. Select an arbitrary value $x_1 \in S$ and define the sequence $\{x_k\}$ by the formula $x_{k+1} = T(x_k)$. T maps from S to S, so

$$x_k \in S, \qquad \forall\, k \geq 1. \tag{A.112}$$

The first step in the proof is to show that the sequence $\{x_k\}$ is a Cauchy sequence. We have

$$\|x_{k+1} - x_k\| = \|T(x_k) - T(x_{k-1})\| \tag{A.113}$$

$$\leq \rho\|x_k - x_{k-1}\| \leq \rho^2\|x_{k-1} - x_{k-2}\| \leq \ldots \leq \rho^{k-1}\|x_2 - x_1\|. \tag{A.114}$$

Then it can be written that

$$\|x_{k+r} - x_k\| \leq \|x_{k+r} - x_{k+r-1}\| + \|x_{k+r-1} - x_{k+r-2}\| + \ldots + \|x_{k+1} - x_k\| \tag{A.115}$$

$$\leq \left(\rho^{k+r-2} + \rho^{k+r-3} + \ldots + \rho^{k-1}\right)\|x_2 - x_1\| \tag{A.116}$$

$$\leq \rho^{k-1}\sum_{i=0}^{\infty}\rho^i\|x_2 - x_1\| = \frac{\rho^{k-1}}{1-\rho}\|x_2 - x_1\|. \tag{A.117}$$

The right-hand side of the equation approaches zero as $k \to \infty$. Therefore, it is a Cauchy sequence. Also, as χ is a Banach space, we have

$$x_k \to x^* \in \chi \qquad \text{when} \qquad k \to \infty. \tag{A.118}$$

Also, since S is closed, $x^* \in S$. Now it will be shown that $x^* = T(x^*)$. For any $x_k = T(x_{k-1})$, we have

$$\|x^* - T(x^*)\| \leq \|x^* - x_k\| + \|x_k - T(x^*)\| \leq \|x^* - x_k\| + \rho\|x_{k-1} - x^*\|. \tag{A.119}$$

If a large enough value of k is chosen, the right-hand side of the inequality can be made arbitrarily small. Thus, $\|x^* - T(x^*)\| = 0$, which is the same as $x^* = T(x^*)$. It remains to prove that x^* is the only fixed point of T in S. Suppose x^* and y^* are fixed points. Then

$$\|x^* - y^*\| = \|T(x^*) - T(y^*)\| \leq \rho\|x^* - y^*\| \tag{A.120}$$

with $\rho < 1$. We obtain

$$x^* = y^*. \tag{A.121}$$

Kinematics and dynamics background

B

This appendix presents a brief background on robots' kinematic and dynamic modeling.

B.1 Kinematics

This section describes how the position and orientation can be represented. The section begins with the frame declaration. After, the position, velocity, and acceleration vectors are commented. The matrix rotation means the orientation between frames. Finally, some interesting properties of the rotation matrix are explained. Assume an orthogonal inertial frame Σ_E named *world frame* and a second orthogonal frame Σ_B attached to a robot. A vector $\boldsymbol{p} \in \mathbb{R}^3$ represents the robot position and it is obtained by measuring the coordinates of the Σ_B origin from Σ_E (see Fig. B.1):

$$\boldsymbol{p} \triangleq \begin{bmatrix} x \\ y \\ z \end{bmatrix}. \tag{B.1}$$

The first and second time derivatives of the position are the velocity and acceleration, denoted by $\dot{\boldsymbol{p}}$ and $\ddot{\boldsymbol{p}}$, respectively:

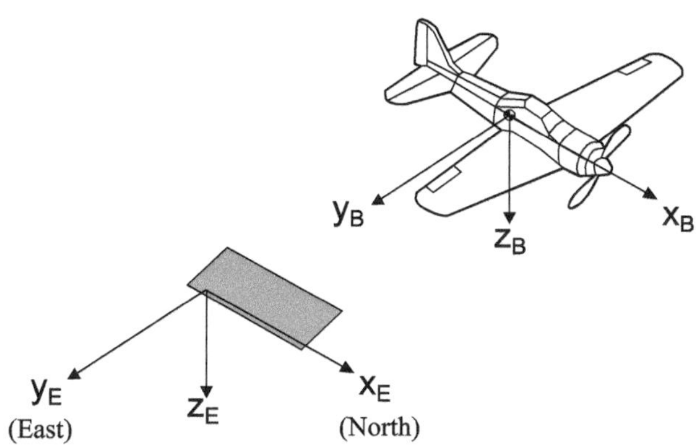

Figure B.1 Free body diagram.

$$\dot{p} = \frac{dp}{dt} = \begin{bmatrix} \dot{x} \\ \dot{y} \\ \dot{z} \end{bmatrix}, \quad \ddot{p} = \frac{d^2 p}{dt^2} = \begin{bmatrix} \ddot{x} \\ \ddot{y} \\ \ddot{z} \end{bmatrix}. \tag{B.2}$$

The matrix $R_B^E \in \mathbb{R}^{3\times 3}$ describes the Σ_B orientation viewed from Σ_E. Let ϕ, θ, and ψ be roll, pitch, and yaw, which are the robot rotation movements, respectively. The matrix R_B^E is obtained as follows:

$$R_B^E(\phi,\theta,\psi) = \begin{bmatrix} \cos(\theta)\cos(\psi) & -\cos(\theta)\sin(\psi) & \sin(\theta) \\ \sin(\phi)\sin(\theta)\cos(\psi)+\cos(\phi)\sin(\phi) & -\sin(\phi)\sin(\theta)\sin(\psi)+\cos(\phi)\cos(\psi) & -\sin(\phi)\cos(\theta) \\ -\cos(\phi)\sin(\theta)\cos(\psi)+\sin(\phi)\sin(\psi) & \cos(\phi)\sin(\theta)\sin(\psi)+\sin(\phi)\cos(\psi) & \cos(\phi)\cos(\theta) \end{bmatrix}. \tag{B.3}$$

The rotation matrices belong to the special group $SO(3)$ and therefore have the following properties.

Property B.1. *The determinant of a rotation matrix is always equal to* 1*:*

$$\det\left(R_B^E(\phi,\theta,\psi)\right) = 1 \qquad \forall \qquad \phi,\theta,\psi \in \mathbb{R}. \tag{B.4}$$

Each element of a rotation matrix is a polynomial with sine and cosine functions. The result of computing the determinant is only a polynomial of sine and cosine functions. Finally, the determinant is rewritten using the following trigonometric identity to simplify the result to value 1:

$$\cos^2(\cdot) + \sin^2(\cdot) = 1. \tag{B.5}$$

Property B.2. *The Euclidean norm of a vector holds the same value after being multiplied by the rotation matrix, i.e.,*

$$p_1 = R_B^E(\phi,\theta,\psi)p_0. \tag{B.6}$$

The rotation matrix is also known as the *direction cosine matrix*. Each row and column of the matrix are unitary vectors. Thus, multiplying a rotation matrix by a vector does not modify the Euclidean norm of the vector.

Property B.3. *The inverse matrix is equal to the transpose matrix. Moreover, it is equivalent to exchanging the direction of rotation:*

$$\mathrm{inv}(R_B^E(\phi,\theta,\psi)) = \left(R_B^E(\phi,\theta,\psi)\right)^{-1} = \left(R_B^E(\phi,\theta,\psi)\right)^\top = R_E^B(\phi,\theta,\psi). \tag{B.7}$$

The inverse matrix can be obtained with the following expression:

$$\mathrm{inv}\left(R_B^E(\phi,\theta,\psi)\right) = \frac{\mathrm{cof}\left(R_B^E(\phi,\theta,\psi)\right)^\top}{\det\left(R_B^E(\phi,\theta,\psi)\right)}, \tag{B.8}$$

where $\text{cof}\left(R_B^E(\phi,\theta,\psi)\right)$ is the cofactor matrix of the rotation matrix. By Property B.1 the expression is rewritten as follows:

$$\text{inv}\left(R_B^E(\phi,\theta,\psi)\right) = \text{cof}\left(R_B^E(\phi,\theta,\psi)\right)^\top. \tag{B.9}$$

It is easy to demonstrate that the cofactor matrix is equal to the rotation matrix using the identity (B.5). Therefore, the final expression is

$$\text{inv}\left(R_B^E(\phi,\theta,\psi)\right) = R_B^E(\phi,\theta,\psi)^\top. \tag{B.10}$$

B.2 Dynamics

In this section, the Euler–Lagrange equation movements are presented. First, the potential and kinetic energies are commented on. After, we described how to use the potential and kinetic energies to obtain the Lagrangian. Finally, the movement equations are obtained with partial and time derivatives of the Lagrangian. Some particular cases are commented on.

The potential energy is obtained with an expression that depends on the mass, gravity, and height with respect to the *world frame*. The potential energy expression is

$$\mathcal{U} = mgh. \tag{B.11}$$

In (B.11) one must be aware of the sign of g. Usually, the *world frame* is oriented in such a way that the gravity and the corresponding axis are parallel. If the gravity has the same direction, the corresponding axis is positive; otherwise the gravity is negative.

Subsequently, the kinetic energy must be obtained. The kinetic energy depends on mass, velocity, inertia moments, and angular velocity. The expression of kinetic energy is

$$\mathcal{K} = \frac{1}{2}m\dot{\boldsymbol{p}}^\top\dot{\boldsymbol{p}} + \frac{1}{2}\boldsymbol{\omega}^\top R_B^E(\phi,\theta,\psi)\mathcal{I}R_E^B(\phi,\theta,\psi)\boldsymbol{\omega}, \tag{B.12}$$

where $\boldsymbol{\omega} \in \mathbb{R}^3$ is the angular velocity and $\mathbb{I} \in \mathbb{R}^3$ is the inertia tensor with respect to the mass center. The inertia tensor is a second-order symmetric tensor, and its elements depend on the mass and the robot geometry.

The values of kinetic and potential energy are used to obtain the Lagrangian as seen in the following expression:

$$\mathcal{L} = \mathcal{K} - \mathcal{U}. \tag{B.13}$$

The Lagrangian is a scalar function with units in Joules. After, $\mathcal{L}$ is used in the Euler–Lagrange movement equations to obtain the linear and angular forces involved in the robot's motion. The movement equations are as follows:

$$\boldsymbol{\tau} = \frac{d}{dt}\left(\frac{\partial \mathcal{L}}{d\dot{\boldsymbol{q}}}\right) - \frac{\partial \mathcal{L}}{\partial \boldsymbol{q}}, \tag{B.14}$$

where $\boldsymbol{\tau}$ is the vector force made up of the linear and angular forces and $\boldsymbol{q}$ and $\boldsymbol{q}$ are vectors that indicate the position and velocity of the *degrees of freedom*. The particular case of aerial robots, the *degrees of freedom* are $\boldsymbol{p}$, ϕ, θ, and ψ. Therefore, the $\boldsymbol{q}$ and $\dot{\boldsymbol{q}}$ vectors are defined as follows:

$$\boldsymbol{q} = \begin{bmatrix} x & y & z & \phi & \theta & \psi \end{bmatrix}^{\top}, \qquad \dot{\boldsymbol{q}} = \begin{bmatrix} \dot{x} & \dot{y} & \dot{z} & \dot{\phi} & \dot{\theta} & \dot{\psi} \end{bmatrix}^{\top}. \tag{B.15}$$

After solving the derivatives of (B.14) the expression must have the following structure:

$$\boldsymbol{\tau} = M(q)\ddot{\boldsymbol{q}} + C(\boldsymbol{q},\dot{\boldsymbol{q}})\dot{\boldsymbol{q}} + \boldsymbol{g}(\boldsymbol{q}), \tag{B.16}$$

where $M(\boldsymbol{q})$ is the symmetric positive definite matrix, $C(\boldsymbol{q},\dot{\boldsymbol{q}})$ is a square matrix of Coriolis, centrifugal, and gyroscopic forces, and $\boldsymbol{g}(\boldsymbol{q})$ is the gravity force vector. The matrices $M(\boldsymbol{q})$ and $C(\boldsymbol{q},\dot{\boldsymbol{q}})$ can be separated as follows:

$$M(\boldsymbol{q}) = M_1(\boldsymbol{q}) + M_2, \tag{B.17}$$

$$C(\boldsymbol{q}) = C_1(\boldsymbol{q},\dot{\boldsymbol{q}}) + C_2(\dot{\boldsymbol{q}}). \tag{B.18}$$

The matrix $M_1(\boldsymbol{q})$ is a symmetric positive definite matrix, containing the terms that depend on $\boldsymbol{q}$. The matrix M_2 is also symmetric positive definite but it is constant. $C_1(\boldsymbol{q},\dot{\boldsymbol{q}})$ is the matrix of Coriolis and centrifugal forces, and $C_2(\dot{\boldsymbol{q}})$ is a skew-symmetric matrix that represents the gyroscopic forces. It is not necessary that the four matrix exist, but some relations between them exist; for instance, if $M_1(\boldsymbol{q})$ exists, $C_1(\boldsymbol{q},\dot{\boldsymbol{q}})$ also exists, and the same relation holds between M_2 and $C_2(\dot{\boldsymbol{q}})$. The following equality is also satisfied:

$$\boldsymbol{q}^{\top}\left[\frac{1}{2}\frac{d}{dt}M(\boldsymbol{q}) - \frac{1}{2}C(\boldsymbol{q},\dot{\boldsymbol{q}})\right]\boldsymbol{q} = 0. \tag{B.19}$$

Stability in the Lyapunov sense

C

The *stability* in dynamical systems is a binary property; this means that a dynamical system is stable or not stable, only one of both. In the literature different criteria to establish stability can be found, like *Lagrange stability*, *bounded input bounded output*, Routh, and Hurwitz, to name a few. This section presents a brief background for stability in the Lyapunov sense. For a more in-depth study of stability, it is recommended to review the reference [38].

In the literature two methods proposed by Lyapunov can be found. The first method requires knowledge of the dynamical system time response. The second method is named the *direct Lyapunov method*, and it does not require a time response. In the following, only the *direct Lyapunov method* is presented.

Assume $\boldsymbol{x}(t) \in \mathbb{R}^n$ is the state vector and consider a dynamical model with the form

$$\dot{\boldsymbol{x}}(t) = \boldsymbol{f}(\boldsymbol{x}(t)) \quad (t > t_0). \tag{C.1}$$

The function $\boldsymbol{f} : \mathbb{R}^n \to \mathbb{R}^n$ is a nonlinear autonomous vector function. The first step to establish if a dynamical system is *stable* is finding the equilibrium points of the system. The equilibrium points are the values of $\boldsymbol{x}$ that satisfy the equation $\boldsymbol{f}(\boldsymbol{x}) = \boldsymbol{0}$. The stability theory for linear cases could be resumed with the following definitions.

Definition C.1 (Free system). The dynamical system (C.1) is called a *free system*. It has the form

$$\dot{\boldsymbol{x}}(t) = A(t)\boldsymbol{x}(t) + B(t)\boldsymbol{u}(\boldsymbol{x}(\boldsymbol{t})). \tag{C.2}$$

Definition C.2 (Free system time response). The time response of (C.2) can be written with transition matrix $\Phi(t, t_0)$ with t_0 as initial time as follows:

$$\boldsymbol{x}(t) = \Phi(t, t_0) + \int_{t_0}^{t} \Phi(\tau, t_0)B(\tau)\boldsymbol{u}(\boldsymbol{x}(\tau))d\tau. \tag{C.3}$$

Definition C.3 (L-stability). The equilibrium point $\boldsymbol{x} = 0$ of (C.2) is called *L-stable* (i.e., stable in the Lyapunov sense) if for every real number $\epsilon > 0$ and every initial time t_0, there exists some number δ as small as desired that depends on t_0 and ϵ such that if $\|\boldsymbol{x}_0\| < \delta$, then $\|\boldsymbol{x}_0\| < \epsilon \; \forall \, t \geq t_0$.

Definition C.4 (Asymptotic stability). The equilibrium point $\boldsymbol{x} = \boldsymbol{0}$ is asymptotically stable if:

- it is *L-stable*,
- for every $\boldsymbol{x_0}$ and $\boldsymbol{t_0}$ sufficiently close to $\boldsymbol{x} = \boldsymbol{0}$, the condition $\boldsymbol{x}(t) \to \boldsymbol{0}$ for $t \to \infty$ holds.

Definition C.5 (Uniform L-stability). If the value of δ does not depend on t_0, then the system (C.2) has uniform L-stability.

Definition C.6 (Uniform asymptotic stability). The system (C.2) is uniform asymptotic stable if and only if two constants $k_1, k_2 > 0$ such that $\|\boldsymbol{\Phi}(t, t_0)\| \leq k_1 e^{-k_2(t-t0)}$ exist.

Definition C.7 (Lyapunov candidate function (LCF)). Let $V : \mathbb{R} \rightarrow \mathbb{R}_+$ be a scalar continuous and differentiable function. It is an LCF for the equilibrium point $\boldsymbol{x} = \mathbf{0}$ if the following conditions are satisfied:

- $V(\boldsymbol{x}(t))$ is locally or globally positive definite,
- $\dfrac{\partial V(\boldsymbol{x}(t))}{\partial \boldsymbol{x}}$ is continuous with respect to $\boldsymbol{x}$.

Definition C.8 (LCF time derivative). The total time derivative of $V(\boldsymbol{x}(t))$ along the trajectories of (C.1) denoted by $\dot{V}(\boldsymbol{x}(t))$ is given by

$$\frac{dV(\boldsymbol{x}(t))}{dt} = \dot{V}(\boldsymbol{x}(t)) = \frac{\partial V(\boldsymbol{x}(t))}{\partial \boldsymbol{x}} \boldsymbol{f}(\boldsymbol{x}). \tag{C.4}$$

Definition C.9 (Lyapunov function). The LCF $V(\boldsymbol{x}(t))$ is called the *Lyapunov function* of (C.1) for small $\|\boldsymbol{x}\|$ if the total time derivative satisfies

$$\dot{V}(\boldsymbol{x}(t)) \leq 0 \qquad \forall \qquad t \geq 0. \tag{C.5}$$

C.1 Direct Lyapunov method

Let $d(\boldsymbol{x}(t), 0)$ be a scalar function that computes the distance between the state $\boldsymbol{x}(t)$ and the origin $\boldsymbol{x} = 0$ using any valid norm. If $d(\boldsymbol{x}(t), 0) \rightarrow 0$ for $t \rightarrow \infty$ and the origin is an equilibrium point, then the origin is an asymptotically stable equilibrium point. Nevertheless, it is not necessary to compute the distance to the concluded asymptotic stability of a dynamical system with the direct Lyapunov method. Instead, an energy function named the *Lyapunov function* is used for such purpose.

Theorem 1 (Stability). *The equilibrium point $\boldsymbol{x} = \mathbf{0}$ is stable if there exists a Lyapunov function for the dynamical system* (C.1).

Theorem 2 (Uniform stability). *If the equilibrium point $\boldsymbol{x} = \mathbf{0}$ satisfies Eq.* (C.1) *and $V(\boldsymbol{x}(t))$ is decreasing for small $\|\boldsymbol{x}\|$, then the equilibrium is uniformly stable.*

For Theorems 1 and 2 we conclude that the stability is local if the LCF used is locally positive definite.

Theorem 3 (Global asymptotic stability). *If the equilibrium point $\boldsymbol{x} = \mathbf{0}$ of* (C.1) *is the only equilibrium point and the LCF is radially unbounded and globally positive definite, then the equilibrium is globally asymptotic stable if the total time derivative of $V(\boldsymbol{x}(t))$ satisfies*

$$\dot{V}(\boldsymbol{x}(t)) < 0 \qquad \forall \qquad \boldsymbol{x}(t) \neq \mathbf{0}. \tag{C.6}$$

Theorem 4 (La Salle). *Assume that the LCF is radially unbounded and globally positive definite, but its time derivative is negative semidefinite:*

$$\dot{V}(\boldsymbol{x}(t)) \leq 0 \qquad \forall \qquad \boldsymbol{x} \in \mathbb{R}^n. \tag{C.7}$$

It is possible to define a set $\Omega = \left\{\boldsymbol{x} \in \mathbb{R}^n : \dot{V}(\boldsymbol{x}) = 0\right\}$. *If the origin (i.e.,* $\boldsymbol{x} = \mathbf{0}$*) is the only equilibrium point and it is contained in* Ω*, then the origin is globally asymptotic stable.*

Fundamentals of linear and nonlinear controllers

D

In Sections D.1 and D.2 basic theory used in this book about linear and nonlinear controllers is presented, respectively. Students and researchers are stimulated to review other texts about these topics as well.

D.1 Fundamentals of linear controllers

D.1.1 PID and PD controller theory

The PID controller is a combination of the proportional (P), integral (I), and derivative (D) control actions. It produces a control signal proportional to the error signal, which is the input to the controller and is equal to the difference between the variable to be controlled and its corresponding desired reference.

The PID controller is considered the most straightforward design, but it may or may not be very effective depending on the application. The proportional controller is essentially a gain amplifier.

In the integral action (I), the control signal is proportional to the integral of the error, so it could also be considered a proportional type controller (P) with a gain that increases as the error persists. This property is often needed to achieve zero steady-state output error. Derivative control (D) produces a control signal proportional to the error change rate. Derivative control includes damping in the system and is typically applied in conjunction with other control elements.

Since derivative control alone cannot drive the error to zero, it is necessary to complement it with a proportional action. The general PID control equation is [65]

$$u(t) = K_p e(t) + K_i \int_0^t e(t)dt + K_v \frac{de(t)}{dt}, \tag{D.1}$$

where $e(t)$ is the error signal, K_p represents the proportional gain, K_i denotes the integral gain, and K_v corresponds to derived profit. Here $u(t)$ is the control signal applied to the plant (system to be controlled).

The PD controller is a particular case of the PID controller in which the integral action is not included. The PD equation can be written as

$$u(t) = K_p e(t) + K_v \frac{de(t)}{dt}. \tag{D.2}$$

Fig. D.1 shows a PID controller applied to a plant. Suppose a mathematical model of the plant can be obtained. In that case, various design techniques and gain tuning

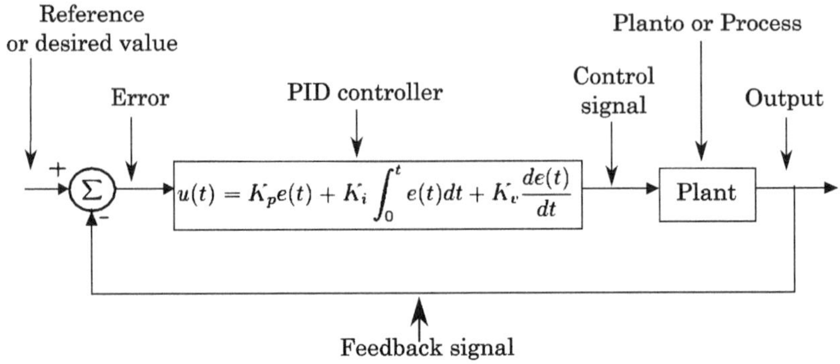

Figure D.1 Block diagram of a PID controller in a plant.

can be used to determine the controller parameters that meet the transient and steady-state specifications of the closed-loop system. However, if the plant is so complicated that its mathematical model is not easy to obtain, an analytical method for designing a PID controller is also complicated. In this case, experimental procedures must be used to tune the PID controllers. Particular cases of the said controller can be achieved with each action that forms it; one of them is the PD controller.

D.1.2 Linear quadratic regulator (LQR)

The design of the LQR is necessary to define a system of the form

$$\dot{\mathbf{x}} = \mathbf{A}\mathbf{x} + \mathbf{B}\mathbf{u}. \tag{D.3}$$

In this controller $\mathbf{K}$ defines the optimal vector, and the controller is defined as

$$\mathbf{u}(t) = -\mathbf{K}\mathbf{x}(t), \tag{D.4}$$

which minimizes the cost function

$$J = \int_0^\infty (\mathbf{x}^*\mathbf{Q}\mathbf{x} + \mathbf{u}^*\mathbf{R}\mathbf{u})dt. \tag{D.5}$$

The matrices which determine the relative importance of the error and the cost of energy are $\mathbf{Q}$ and $\mathbf{R}$, where $\mathbf{Q}$ is a positive definite (or positive semidefinite) Hermitian matrix and $\mathbf{R}$ is a positive definite Hermitian matrix; a Hermitian matrix is a complex square matrix that is equal to its conjugate transpose. In this case, if the control vector $\mathbf{u}(t)$ is not restricted, then (D.4) is the optimal control law obtained. Therefore, if the unknown elements of the matrix $\mathbf{K}$ are determined to minimize the cost function, then $\mathbf{u}(t) = -\mathbf{K}\mathbf{x}(t)$ is optimal for any initial state $\mathbf{x}(0)$. Now the optimization problem is solved. Substituting Eq. (D.4) into Eq. (D.3), we get

$$\dot{\mathbf{x}} = \mathbf{A}\mathbf{x} - \mathbf{B}\mathbf{K}\mathbf{x} = (\mathbf{A} - \mathbf{B}\mathbf{K})\mathbf{x}. \tag{D.6}$$

In the following deductions, it is assumed that the matrix $\mathbf{A} - \mathbf{BK}$ is stable or that the eigenvalues of $\mathbf{A} - \mathbf{BK}$ have negative real parts. Substituting Eq. (D.4) into Eq. (D.5) results in

$$J = \int_0^\infty (\mathbf{x}^*\mathbf{Q}\mathbf{x} + \mathbf{x}^*\mathbf{K}^*\mathbf{R}\mathbf{K}\mathbf{x})dt \tag{D.7}$$

$$= \int_0^\infty \mathbf{x}^*(\mathbf{Q} + \mathbf{K}^*\mathbf{R}\mathbf{K})\mathbf{x}dt. \tag{D.8}$$

Then

$$\mathbf{x}^*(\mathbf{Q}+\mathbf{K}^*\mathbf{R}\mathbf{K})\mathbf{x} = -\dot{\mathbf{x}}^*\mathbf{P}\mathbf{x} - \mathbf{x}^*\mathbf{P}\mathbf{x} - \mathbf{x}^*\mathbf{P}\dot{\mathbf{x}} = \mathbf{x}^*[(\mathbf{A}-\mathbf{BK})^*\mathbf{P}+\mathbf{P}(\mathbf{A}-\mathbf{BK})]\mathbf{x}. \tag{D.9}$$

When comparing both sides of (D.9) and considering that it has to be valid for any $\mathbf{x}$, it is required that

$$(\mathbf{A} - \mathbf{BK})^*\mathbf{P} + \mathbf{P}(\mathbf{A} - \mathbf{BK}) = -(\mathbf{Q} + \mathbf{K}^*\mathbf{R}\mathbf{K}). \tag{D.10}$$

It can be shown that if **A-BK** is a stable matrix, there exists a positive definite matrix **P** that satisfies formula (D.10). Therefore, the procedure is based on determining the elements of **P** from formula (D.10) and seeing if it is positive definite. Note that more than one matrix **P** can satisfy this condition; if the system is stable, there is always a positive definite matrix **P** that satisfies this formula. This means that if the formula is solved and a positive definite matrix **P** is found, the system is stable. Other matrices **P** that satisfy this equivalence are not positive definite and should be discarded. The behavior index J is calculated as

$$J = \int_0^\infty \mathbf{x}^*(\mathbf{Q}+\mathbf{K}^*\mathbf{R}\mathbf{K})\mathbf{x}dt = -\mathbf{x}^*\mathbf{P}\mathbf{x}|_0^\infty = -\mathbf{x}^*(\infty)\mathbf{P}\mathbf{x}(\infty) + \mathbf{x}^*(0)\mathbf{P}\mathbf{x}(0). \tag{D.11}$$

Since all eigenvalues of **A-BK** are assumed to have negative real parts, we have $\mathbf{x}(\infty) \to 0$. Therefore, we obtain

$$J = \mathbf{x}^*(0)\mathbf{P}\mathbf{x}(0). \tag{D.12}$$

Thus, the behavior index J is obtained in terms of the initial condition $x(0)$ and **P**. To obtain the solution to the quadratic optimal control problem, we proceed as follows. Since it has been assumed that **R** is a real symmetric or positive definite Hermitian matrix, we can write

$$\mathbf{R} = \mathbf{T}^*\mathbf{T}, \tag{D.13}$$

where **T** is a nonsingular matrix. Thus, Eq. (D.10) is written as

$$(\mathbf{A}^* - \mathbf{K}^*\mathbf{B}^*)\mathbf{P} + \mathbf{P}(\mathbf{A}\text{-}\mathbf{BK}) + \mathbf{Q} + \mathbf{K}^*\mathbf{T}^*\mathbf{T}\mathbf{K} = 0, \tag{D.14}$$

which can be rewritten as

$$\mathbf{A}^*\mathbf{P}+\mathbf{P}\mathbf{A}+[\mathbf{T}\mathbf{K} - (\mathbf{T}^*)^{-1}\mathbf{B}^*\mathbf{P}]^*[\mathbf{T}\mathbf{K} - (\mathbf{T}^*)^{-1}\mathbf{B}^*\mathbf{P}] - \mathbf{P}\mathbf{B}\mathbf{R}^{-1}\mathbf{B}^*\mathbf{P}+\mathbf{Q} = 0. \tag{D.15}$$

The minimization of J with respect to $\mathbf{K}$ requires the minimization of

$$\mathbf{x}^*[\mathbf{TK} - (\mathbf{T}^*)^{-1}\mathbf{B}^*\mathbf{P}]^*[\mathbf{TK} - (\mathbf{T}^*)^{-1}\mathbf{B}^*\mathbf{P}]\mathbf{x} \tag{D.16}$$

with respect to K. Since this last expression is not negative, the minimum occurs when it is zero, or when

$$\mathbf{TK} = (\mathbf{T}^*)^{-1}\mathbf{B}^*\mathbf{P}. \tag{D.17}$$

Thus

$$\mathbf{K} = \mathbf{T}^{-1}(\mathbf{T}^*)^{-1}\mathbf{B}^*\mathbf{P} = \mathbf{R}^{-1}\mathbf{B}^*\mathbf{P}. \tag{D.18}$$

Eq. (D.18) gives the optimal matrix K. Thus, the optimal control law for the quadratic optimal control problem is linear when the behavior index is given by Eq. (D.5) and can be expressed as

$$\mathbf{u}(t) = -\mathbf{Kx}(t) = -\mathbf{R}^{-1}\mathbf{B}^*\mathbf{Px}(t). \tag{D.19}$$

The $\mathbf{P}$ matrix in Eq. (D.18) must satisfy Eq. (D.10) or the following reduced equation:

$$\mathbf{A}^*\mathbf{P} + \mathbf{PA} - \mathbf{PBR}^{-1}\mathbf{B}^*\mathbf{P} + \mathbf{Q} = 0. \tag{D.20}$$

Eq. (D.20) is called the reduced Riccati matrix equation. Fig. D.2 presents an optimal regulatory system. The design steps are outlined as follows:

1. Solve formula (D.20), which is the reduced Riccati formula, for the matrix $\mathbf{P}$, that is, if there exists a positive definite matrix $\mathbf{P}$.
2. Substitute this matrix $\mathbf{P}$ into Eq. (D.18). Then, matrix $\mathbf{K}$ turns out to be the optimal one.

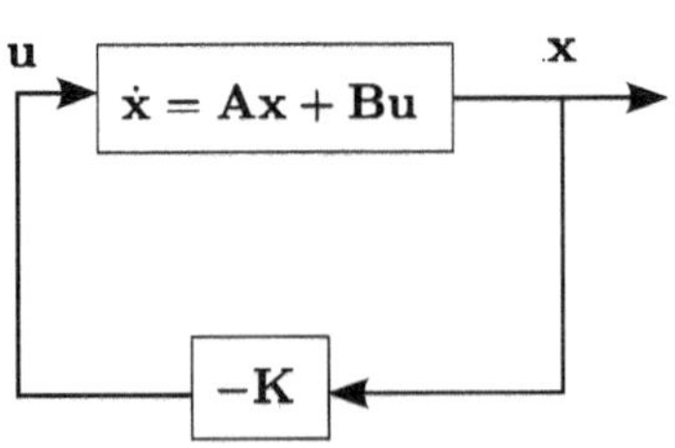

Figure D.2 Optimal regulatory system.

D.2 Fundamentals of nonlinear controllers

D.2.1 Nested saturations

The objective of the nested saturation controller involves the stabilization of a chain of integrators [63]:

$$\dot{x}_1 = x_2, \tag{D.21}$$

$$\dot{x}_2 = x_3, \tag{D.22}$$

$$\vdots \tag{D.23}$$

$$\dot{x}_{n-1} = x_n, \tag{D.24}$$

$$\dot{x}_n = u. \tag{D.25}$$

This chain of integrators can be represented as a state space. For the state equation $\dot{x} = A_x x + B_x u$ with $x \in \mathbb{R}^n$ and $u \in \mathbb{R}$, the matrices A_x and B_x are defined as

$$A_x = \begin{pmatrix} 0 & 1 & 0 & \cdots & 0 \\ \vdots & & & & \vdots \\ 0 & \cdots & \cdots & \cdots & 1 \\ 0 & \cdots & \cdots & \cdots & 0 \end{pmatrix}, \tag{D.26}$$

$$B_x = \begin{pmatrix} 0 \\ \vdots \\ \vdots \\ 1 \end{pmatrix}. \tag{D.27}$$

Starting from these equations, we consider if there exists a linear transformation defined as $z = T_{zx}x$. Then (D.21) is transformed in $\dot{z} = A_z z + B_z u$, and the matrices A_z and B_z are defined as

$$A_z = \begin{pmatrix} 0 & k_2 & k_3 & \cdots & k_{n-1} & k_n \\ 0 & 0 & k_3 & \cdots & k_{n-1} & k_n \\ \vdots & \ddots & \ddots & \ddots & \vdots & \vdots \\ 0 & \cdots & \cdots & \cdots & 0 & k_n \\ 0 & \cdots & \cdots & \cdots & 0 & 0 \end{pmatrix}, \tag{D.28}$$

$$B_z = \begin{pmatrix} 1 \\ 1 \\ \vdots \\ \vdots \\ 1 \end{pmatrix}, \tag{D.29}$$

where $k_1, k_2, ..., k_{n-1}, k_n$ are called coefficients in order to achieve the transformation, which belong to $\mathbb{R}$. Now we consider the coefficient set $K = k_n, k_{n-1}, ..., k_2, k_1$. Let $A_l \subseteq A$ represent a subset containing the first l elements of A. It is necessary to define a function $F_k^m(A_l)$ which is acting in the set A_l; F_k^m generates the product of combinations of elements considering m at a time from A_l. The binomial coefficient $C_m^l = \binom{l}{m}$ is used to the number of such combinations. With the aim to generate a transformation T_{zx} we define a function $C(m, l)$ with $l \in [0, ..., n]$, $m \in [0, ..., l]$, and $m \lneq l$ over the set of coefficients of A defined by $K = k_n, k_{n-1}, ..., k_2, k_1$. Then we have

$$C(l, m) = \sum_{h=1}^{C_m^l} F_h^m(K_l), \tag{D.30}$$

$$C(l, 0) = 1. \tag{D.31}$$

Thus, the system in the new coordinates is given by

$$z_{n-i} = \sum_{j=0}^{i} C(i, j) x_{n-j}, \tag{D.32}$$

with $i \in [0, ..., n-1]$, and the transformation T_{zx} is defined by

$$T_{zx_{(n-i)(n-j)}} = C(i, j), \qquad i \geq j, \tag{D.33}$$

$$T_{zx_{(n-i)(n-j)}} = 0, \qquad i < j. \tag{D.34}$$

After introducing the aforementioned, a variable transformation of a system composed by two integers in cascade is obtained:

$$K = k_2, k_1, \qquad k_1 = k_2, \tag{D.35}$$

$$F_1^1 = k_2, \qquad F_1^0 = 1. \tag{D.36}$$

Also,

$$C(0, 0) = 1, \tag{D.37}$$

$$C(1, 0) = F_1^0(k_1) = 1, \tag{D.38}$$

$$C(1, 1) = F_1^1(k_1) = k_2. \tag{D.39}$$

After this, a transformation matrix of two integrators is obtained of the following form:

$$\begin{pmatrix} z_1 \\ z_2 \end{pmatrix} = \begin{pmatrix} T_{zx1,1} & T_{zx1,2} \\ 0 & T_{zx1,2} \end{pmatrix} \begin{pmatrix} x_1 \\ x_2 \end{pmatrix}, \tag{D.40}$$

with

$$T_{zx2,2} = C(0, 0) = 1, \tag{D.41}$$

$$T_{zx1,2} = C(1,0) = F_1^0(k1) = 1, \tag{D.42}$$
$$T_{zx2,1} = C(0,1) = 0, \tag{D.43}$$
$$T_{zx1,1} = C(1,1) = F_1^1(k_1) = k_2. \tag{D.44}$$

Finally, the bounded control which results in an asymptotically stable system for a chain of two integrators is given by $u = -\sigma_n(z_n + \sigma_{n-1}(z_{n-1}))$ with $n = 2$.

D.2.2 Integrator backstepping

Integrator backstepping is a nonlinear control theory. To explain it in a general form, let us define a system of the following form:

$$\dot{\varphi} = f(\varphi) + g(\varphi)\varrho, \tag{D.45}$$
$$\dot{\varrho} = u, \tag{D.46}$$

where the state variables are defined by φ and ϱ, which belong to the Euclidean space $\mathbb{R}^n$, u defines the control input, and $u \in \mathbb{R}$, f, and g are smooth functions. Thus, the objective is to design a state feedback controller to stabilize the origin, that is, $\varphi = 0$ and $\varrho = 0$. It should mentioned that it is assumed that the functions f and g are known. Thus, suppose that (D.45) can be stabilized by a state feedback controller $\varrho = \rho(\varphi)$ and consider $\rho(0) = 0$; this means that the origin of $\dot{\varphi} = f(\varphi) + g(\varphi)\rho(\varphi)$ is asymptotically stable. Suppose that a smooth, positive definite Lyapunov function $V(\varphi)$ is known, and this Lyapunov function satisfies the inequality $\frac{\partial V}{\partial \varphi}[f(\varphi) + g(\varphi)\rho(\varphi)] \leq -W(\varphi)$ with $W(\varphi) > 0$. Then, by adding and subtracting $g(\varphi)\varrho$ on the right-hand side of (D.45), we obtain

$$\dot{\varphi} = [f(\varphi) + g(\varphi)\rho(\varphi)] + g(\varphi)[\varrho - \rho(\varphi)], \tag{D.47}$$
$$\dot{\varrho} = u. \tag{D.48}$$

Then a change of variables is defined as

$$\chi = \varrho - \rho(\varphi). \tag{D.49}$$

Thus, the change of variables is use to the backstep, and after the following system is defined:

$$\dot{\varphi} = [f(\varphi) + g(\varphi)\rho(\varphi)] + g(\varphi)\chi, \tag{D.50}$$
$$\dot{\chi} = u - \dot{\rho}(\varphi). \tag{D.51}$$

The derivative of $\dot{\rho}(\varphi)$ is computed by the expression $\dot{\rho}(\varphi) = \frac{\partial \rho}{\partial \varphi}[f(\varphi) + g(\varphi)\varrho]$. Let us take $v = u - \dot{\rho}(\varphi)$ to transform the system into the following form:

$$\dot{\varphi} = [f(\varphi) + g(\varphi)\rho(\varphi)] + g(\varphi)\chi, \tag{D.52}$$
$$\dot{\chi} = v. \tag{D.53}$$

It can be seen that the system (D.52)–(D.53) is similar to the system (D.45)–(D.46), but now the first component presents an asymptotic stable origin if the input is zero. Let us define an LCF:

$$V_c(\varphi,\varrho)=V(\varphi)+\frac{1}{2}\chi^2. \tag{D.54}$$

Taking the derivative, the LCF is obtained:

$$\dot{V}_c=\frac{\partial V}{\partial \varphi}[f(\varphi)+g(\varphi)\rho(\varphi)]+\frac{\partial V}{\partial \varphi}g(\varphi)\chi+\chi v\leq -W(\varphi)+\frac{\partial V}{\partial \varphi}g(\varphi)\chi+\chi v. \tag{D.55}$$

If we choose a $v=\frac{\partial V}{\partial \varphi}g(\varphi)-k\chi$, with k as a positive definite gain, then the final derivative of (D.55) is given by

$$\dot{V}_c\leq -W(\varphi)-k\chi^2. \tag{D.56}$$

Thus, the origin $\varphi=0$ and $\chi=0$ is asymptotically stable. Since $\rho(0)=0$ it is possible to conclude that the origin $\varphi=0$ and $\varrho=0$ is asymptotically stable, and substituting for v, χ, and $\dot{\rho}$ the state feedback controller is obtained:

$$u=\frac{\partial \rho}{\partial \varphi}[f(\varphi)+g(\varphi)\varrho]-\frac{\partial V}{\partial \varphi}g(\varphi)-k[\varrho-\rho(\varphi)]. \tag{D.57}$$

Then, we conclude that the origin is globally asymptotically stable if and only if all the assumptions hold globally and $V(\varphi)$ is radially unbounded [41], [44].

D.2.3 Sliding mode control

Consider the second-order system given by

$$\dot{x}_1=x_2, \tag{D.58}$$
$$\dot{x}_2=f(x)+g(x)u, \tag{D.59}$$

where $f(x)$ and $g(x)$ are unknown nonlinear functions and $g(x)\geq 0>0$ for all x. It is desired to design a state feedback control law to stabilize the origin. Assume that it is possible to design a control law such that the motion constraints of the system remain within the sliding surface $\sigma_x=k_1x_1+x_2=0$ [59], [41], [60]. For this sliding surface, the motion is governed by $\dot{x}_1=-k_1x_1$. With k_1 as a definite positive gain it is guaranteed that $x(t)\longrightarrow 0$ as $t\longrightarrow\infty$ and the rate of convergence is controlled by the choice of k_1. The motion on the sliding surface $\sigma_x=0$ is independent of $f(x)$ and $g(x)$. The variable σ_x satisfies the equation

$$\dot{\sigma}_x=k_1\dot{x}_1+\dot{x}_2=k_1x_2+f(x)+g(x)u. \tag{D.60}$$

Assume $f(x)$ and $g(x)$ satisfy the inequality

$$\left|\frac{\nu}{g(x)}\right|\leq\varrho(x) \tag{D.61}$$

with $\nu = k_1 x_2 + f(x)$, $\forall x \in \mathbb{R}^2$, and for a known function $\varrho(x)$. Define a $V = \frac{1}{2}\sigma_x^2$ as an LCF for $\dot{\sigma}_x = k_1 x_2 + f(x) + g(x)u$. Then we compute the derivative of the LCF,

$$\dot{V} = \sigma_x \dot{\sigma}_x = \sigma_x [k_1 x_2 - f(x)] + g(x)\sigma_x u \le g(x)|\sigma_x|\varrho(x) + g(x)\sigma_x u, \quad \text{(D.62)}$$

and define the sliding mode control as

$$u = -\beta(x)\,\mathrm{sgn}(\sigma_x). \quad \text{(D.63)}$$

It can be seen that $u = -\beta_x\,\mathrm{sgn}(\sigma_x)$ is used only for $\sigma_x \neq 0$ since in ideal sliding mode control u is undefined on the sliding surface $\sigma_x = 0$. Alternatively, $u = -\beta_x\,\mathrm{sgn}(\sigma_x)$ can be written for all σ_x if $\mathrm{sgn}(\sigma_x)$ is undefined over $\sigma_x = 0$. The following observation applies along the ideal sliding mode control, where $\beta_x \ge \varrho(x) + \beta_0$:

$$\mathrm{sgn}(\sigma_x) = \begin{cases} 1, & \sigma_x > 0, \\ 0, & \sigma_x = 0, \\ -1, & \sigma_x < 0. \end{cases} \quad \text{(D.64)}$$

The sliding mode controller can be simplified in a domain of interest, having $\nu = k_1 x_2 + f(x)$ and $g(x)$ satisfying the inequality

$$\left| \frac{\nu}{g(x)} \right| \le k_1 \quad \text{(D.65)}$$

for a known nonnegative constant k_1. In this case, you can take

$$u = -\beta_x\,\mathrm{sgn}(\sigma_x) \quad \text{(D.66)}$$

with $\beta_x > k_1$, where definition u takes the form of a simple relay, but if this performance is not good due to the chattering effect generated by the sliding mode control, and this chattering effect produces heating in the machines and mechanical vibrations, an electronic system is necessary which produces a high speed in the control signal to obtain the desired control of the system. These are some problems of slider mode control. A solution to avoid or reduce the chattering effect without using high-order sliding modes is using a saturation function $\mathrm{sat}(\sigma_x)$ or a hyperbolic tangent function $\tanh(\sigma_x)$.

D.2.4 Model reference adaptive control (MRAC)

MRAC was introduced by Whitacker in 1958, in the general scheme presented in Fig. D.3. We see a block called *adaptive controller*, which is the primary controller to obtain the closed-loop performance, the *adjustment mechanism* block, which is the part where the adaptive gains are estimated, the *plant* block, which is the process to control, and the *model reference* block, which contains the equations with the desired performance for the plant [3]. Based on the block diagram of Fig. D.3 the model reference can be considered as

$$\dot{y}_p = A_m x + B_m u, \quad \text{(D.67)}$$

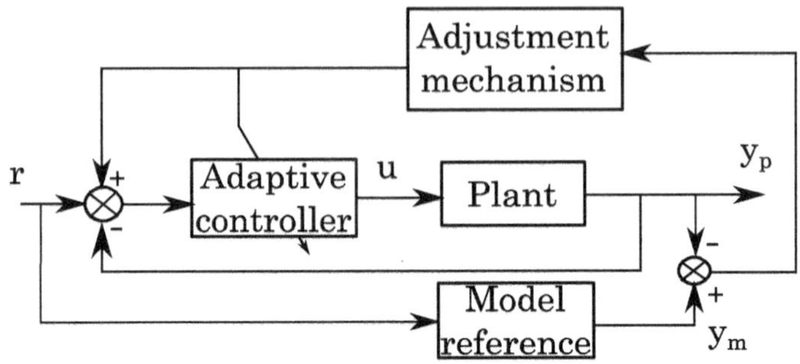

Figure D.3 General block diagram of the MRAC.

where x is the state vector and u is the input vector. It should be mentioned that the model (D.67) is supposed to be a stable and completely controllable system. The plant can be represented as

$$\dot{y} = A_s(e,t)x + B_s(e,t)u, \tag{D.68}$$

with the error defined as $e = y_p - y$. The design objective for this case is to find an adaptive control law such that parameter matrices $A_s(e,t)$ and $B_s(e,t)$ can be modified so that the error tends to zero for any input u. The error derivative can be obtained subtracting the previous equations, where in addition the term $A_m y$ is added and subtracted. Then

$$\dot{e} = A_m x + B_m u - A_s(e,t)x - B_s(e,t)u + A_m y - A_m y. \tag{D.69}$$

Eq. (D.69) can be defined as

$$\dot{e} = A_m e + [A_m - A_s(e,t)]y + [B_m - B_s(e,t)]u. \tag{D.70}$$

Thus considering the error it is possible to obtain an adjustment mechanism in such a way that the loss function is minimized, which is given by

$$J(\theta) = \frac{1}{2}e^2. \tag{D.71}$$

It should be mentioned that it is necessary to make J small to change the parameters in the direction of the negative gradient of J:

$$\frac{d\theta}{dt} = -\gamma\frac{\partial J}{\partial \theta} = -\gamma e\frac{\partial e}{\partial \theta}. \tag{D.72}$$

Eq. (D.72) is the called *MIT rule*. The partial derivative $\frac{\partial e}{\partial \theta}$ is the sensitive derivative of the system, and γ is the adaptation gain.

Discrete-time Kalman filter

E

This appendix presents a brief definition of the discrete-time Kalman filter used in this book to develop the LQR with the discrete Kalman filter to estimate some state variables before and after such estimations are used as feedback for the LQR controller.

The discrete-time Kalman filter used as an optimal estimator of states uses updating equations (prediction) and equations for updating the measure (correction). The equations for the prediction are given by

$$\hat{x}_k^- = A\hat{x}_{k-1} + Bu_k, \tag{E.1}$$

$$P_k^- = AP_{k-1}A^T + Q, \tag{E.2}$$

where $\hat{x}_k^-$ is the prior estimate, A_k is called the transition matrix of the system, B_k is the distribution matrix, u_k is the input vector, P_k^- is the prior covariance error, and Q is the process covariance noise. Then, the equations for the correction are defined as

$$K_k = P_k^- H^T (HP_k^- H^T + R)^{-1}, \tag{E.3}$$

$$\hat{x}_k = \hat{x}_k^- + K_k(z_k - Hx_k^-), \tag{E.4}$$

$$P_k = (1 - K_k H)P_k^-, \tag{E.5}$$

where K_k is the discrete-time Kalman gain, H denotes the measurement matrix, and R_k represents the covariance noise of the measurement. In $\hat{x}_k$ we have the current or obtained estimate, z_k is the measurement value, and P_k denotes the covariance error [27]. We also have

$$\hat{x}_k^- = A\hat{x}_{k-1} + Bu_k + w_k \tag{E.6}$$

$$= x_{k-1} + w_k. \tag{E.7}$$

The measurement value or value to estimate z_k is given by

$$\hat{x}_k = Hx_k + v_k \tag{E.8}$$

$$= x_k + v_k. \tag{E.9}$$

Eqs. (E.6) and (E.8) could be reduced if and only if the estimation is performed on a scalar or constant and not on a matrix. For example, if there is no control signal u_k, the matrix A is equal to one because we are working on a scalar, and we know that the value prior to it will be the same; then, the value for $H = 1$ because the measurement(s) are known to be made up of noise. It is not common to find signals with little or no noise; if this happens the matrix $H \neq 1$. Fig. E.1 presents a discrete-time Kalman filter block diagram with the equations defined in this book.

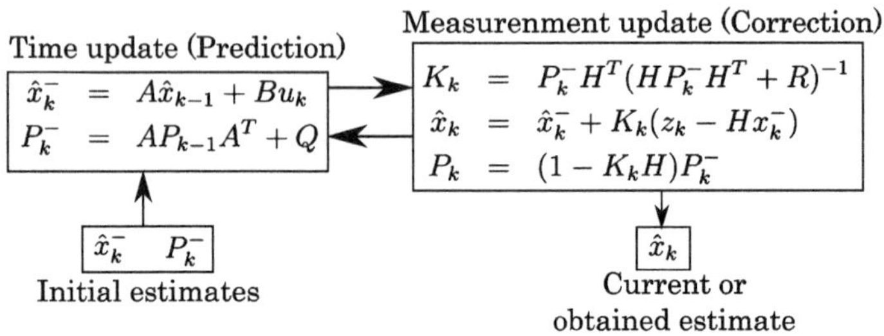

Figure E.1 Discrete-time Kalman filter block diagram.

Linear and nonlinear controllers: programs for the embedded system F

In Section F.1 a program to achieve the desired altitude with the PD controller is presented. The program is written in the Dynamic C language used to program the Rabbit microprocessor.

To understand the programs presented in these appendices it is necessary to know the programming languages C and C++. In Section F.2 the program used in the real flight tests with the nonlinear backstepping technique for altitude is presented.

F.1 PD controller in altitude

```
///IMU libraries///
#class auto
#use "CI3DMGVA4.lib"
/// Declaration servo motors outputs///

#define FAST_INTERRUPT 0
#define Servo_Estabilizador 1  //2 -3
#define Servo_Elevador 2  //3 -1
#define Servo_Alabeo  3  //0 -0

/// Declaration radio signals ///
#define GAS    3
#define YAW    4
#define PITCH  2
#define ROLL   1

///IMU buffer size///
#define BINBUFSIZE  31
#define BOUTBUFSIZE 31

///Xbee buffer size///
#define DINBUFSIZE  255
#define DOUTBUFSIZE 255

/// Altimeter buffer size///
#define CINBUFSIZE  7
#define COUTBUFSIZE 7
```

```
///MATLAB communication buffer size///
char buff_out[56];

/// Structure to read the data from IMU///
typedef struct {
        int i_Roll_rate;
        int i_Pitch_rate;
        int i_Yaw_rate;
        unsigned int u_Roll_rate;
        unsigned int u_Pitch_rate;
        unsigned int u_Yaw_rate;
    float f_Roll_rate;
    float f_Pitch_rate;
    float f_Yaw_rate;
    float Acc_X;
    float Acc_Y;
    float Acc_Z;
    int Mag_X;
    int Mag_Y;
    int Mag_Z;
}Sensores_CI;

/// Structure to read the data from IMU///
typedef struct {
        float pitch;
        float roll;
        float yaw;
        } Attitude;

    float PitchRateGrad,RollRateGrad,YawRateGrad;
    double z;
    double z_des;
    float Pitch_Control,td;

/// Instructions to read IMU///
int obtiene_angulo(Attitude *new_attitude,
Sensores_CI *new_sensor, char *ci_sentence);

///Manual Control///
void input_cap_isr();
int sgn (float x);

///Cursor position///
nodebug void gotoxy (int x, int y)
```

```
{
        x += 0x20;
        y += 0x20;
        printf ("\x1B=%c%c",x,y );
}

///IMU variables///
Attitude  Actual_central;
Attitude Actual_Attitude;
Sensores_CI Actual_sensor;
char Buffer_Central[23];
int CI_state;

///Manual control variables//////
int  PW_Temp , P_Width[8];
char capture_status;
int  x, i, manual;

/// Altimeter functions///
double strtonum(char*);
double z;

///MATLAB functions to send data///
__nodebug
void float2char(float var,int pos)
{

buff_out[pos]  = *((char*)&var);
buff_out[pos+1]= *((char*)&var+1);
buff_out[pos+2]= *((char*)&var+2);
buff_out[pos+3]= *((char*)&var+3);

}
__nodebug
void enviar_datos()
{
int i,checksum;

checksum=0;
buff_out[0]=0xFF;
buff_out[1]=0x02;
float2char(Actual_Attitude.roll ,2);
float2char(Actual_Attitude.pitch ,6);
float2char(Actual_Attitude.yaw,10);
float2char(RollRateGrad ,14);
```

```
float2char(PitchRateGrad ,18);
float2char(YawRateGrad ,22);
float2char(z_des ,26);
float2char(Pitch_Control ,30);
float2char(z,34);
float2char(td ,38);
float2char(0.00 ,42);
float2char(0.00 ,46);
float2char(0.00 ,50);

for(i=2;i<54;i++)
checksum += buff_out[i];

buff_out[54]=*((char*)&checksum);
buff_out[55]=*((char*)&checksum+1);

serDwrite(buff_out, sizeof(buff_out));
}

///Main start///
void main()
{
///Variables declaration///
   int  roll, pitch, yaw;
   int   pwm_options;
   float YawRecta, PitchRecta, RollRecta;
   double T;
   unsigned long T1,t1;
   unsigned long  frecuencia;
   char data[7];
   int Num;
   float e_pitch;
   float pitch_des;
   float kp_p,kv_p;
   float Pitch_Final;
   float Control_PitchRecta;
   int Alt_deseado;
   float e_alt;
   float derivada;
   float e_pitch_anterior;

///Constants declarations///

   roll = 0;
   pitch = 0;
```

```
    yaw = 0;

    Actual_central.roll = 0.0;
    Actual_central.pitch = 0.0;
    Actual_central.yaw = 0.0;

    Actual_sensor.f_Roll_rate = 0.0;
    Actual_sensor.f_Pitch_rate = 0.0;
    Actual_sensor.f_Yaw_rate = 0.0;

    YawRecta = 0.0;
    PitchRecta = 0.0;
    RollRecta = 0.0;

    RollRateGrad=0.0;
    YawRateGrad=0.0;
    PitchRateGrad=0.0;

    T = 0;
    T1 = 0;
    td=0;
    t1=0;

    Alt_deseado=1;
    Pitch_Control=0.0;
    z_des=0.0;

    kp_p = 10;
    kv_p = 0.1;

///Input capture for radio control///

#if __SEPARATE_INST_DATA__ && FAST_INTERRUPT
    interrupt_vector inputcap_intvec input_cap_isr;
#else
    SetVectIntern(0x1A, input_cap_isr);
    SetVectIntern(0x1A, GetVectIntern(0x1A));
#endif

    WrPortI(ICS1R, NULL, 0x44);
    WrPortI(TAT8R, NULL, freq_divider/4-1);
    WrPortI(ICCSR, NULL, 0x1C);
    WrPortI(ICCR, NULL, 0x01);
    WrPortI(ICT1R, NULL, 0x59);
```

```
///PWM configuration///

 frecuencia = pwm_init(495);
 pwm_options = PWM_USEPORTE;
 pwm_options |= PWM_OUTEIGHTH;

///IMU serial port configuration///

    serBopen(115200);
    serBputc(0x06);
    serBread(Buffer_Central ,5 ,5);
    serBrdFlush();
    CI_state = 0;

///Xbee serial port configuration///

     serDopen(115200);
          serDwrFlush();
          serDrdFlush();

/// Altimeter serial port configuration///

     serCopen(38400);
          serCwrFlush();
          serCrdFlush();

/// Central reading cycle start and manual mode///
 while(1)
          {
             costate
      {
wfd  cof_serBputc(0x31);
wfd  CI_state = cof_serBread(Buffer_Central ,23 ,5);
yield;
if (CI_state == 23 && Buffer_Central[0]== 0x31)
{
obtiene_angulo(&Actual_Attitude ,&Actual_sensor ,
                  Buffer_Central);
CI_state = 0;
Buffer_Central[0]=0x00;
}
wfd cof_serDputs(central_datos);
waitfor(DelayMs(90));
}
```

```
  costate {
                serCrdFlush ();
                 wfd  Num = cof_serCread(data ,7 ,100);
                 if(data[6]==0x0B)
                {
                   z=strtonum(data);
                 }
                    }

  costate{
             RollRateGrad=Actual_sensor.f_Roll_rate
                          *57.2958;
             YawRateGrad=Actual_sensor.f_Yaw_rate*57.2958;
             PitchRateGrad=Actual_sensor.f_Pitch_rate
                           *57.2958;
         }

///Servo motors control///
costate{
         YawRecta = 0.4000 * P_Width[YAW] + 180;
           if (YawRecta < 546 )
              {
                YawRecta = 546;
              }
           if (YawRecta > 946)
              {
                YawRecta = 946;
              }
  pwm_set(Servo_Estabilizador , (int)(YawRecta),
           pwm_options);
       }

costate{
  if (P_Width[5] < 1000)
       {
           PitchRecta = 0.4000 * P_Width[PITCH] + 180;
           if (PitchRecta < 546)
              {
                PitchRecta = 546;
              }
           if (PitchRecta > 946)
              {
                PitchRecta = 946;
              }
```

```
        pwm_set(Servo_Elevador, (int)(PitchRecta),
              pwm_options);
           Alt_deseado = 1;
           Pitch_Control = 0.0;
         }
        }

///PD control in altitude definition
costate{
if (P_Width[5]>1000&&CI_state==23&&Buffer_Central[0]
                                    ==0x31)
         {
         if(Alt_deseado==1)
           {
           z_des = z;
           Alt_deseado=0;
           }
            e_alt=z_des-z;
            pitch_des= atan2(z_des-z,14);
           e_pitch= - Actual_Attitude.pitch
                    + deg(pitch_des);

           td = ((double) (MS_TIMER - t1))/1000.0;
           t1 = MS_TIMER;
           derivada = (e_pitch - e_pitch_anterior)/td;
           e_pitch_anterior = e_pitch;

           Pitch_Control = -kp_p*e_pitch - kv_p*derivada;

           Control_PitchRecta = 1 *  Pitch_Control + 746;
            if (Control_PitchRecta < 546)
               {
                 Control_PitchRecta = 546;
               }
            if (Control_PitchRecta > 946)
               {
                 Control_PitchRecta = 946;
               }

           pwm_set(Servo_Elevador,
                 (int)(Control_PitchRecta), pwm_options);
         }
        }

costate{
```

```
            RollRecta = 0.4000 * P_Width[ROLL] + 180;
            if (RollRecta < 546)
               {
                 RollRecta = 546;
               }
            if (RollRecta > 946)
               {
                 RollRecta = 946;
               }

         pwm_set(Servo_Alabeo, (int)(RollRecta),
                 pwm_options);

               }

  costate{
                 serDwrFlush();
                             serDrdFlush();
                 enviar_datos();
                 waitfor(DelayMs(25));
  }
          }
}

///Cofunction definitions///

nodebug int obtiene_angulo(Attitude *new_attitude,
                           Sensores_CI *new_sensor,
                           char *ci_sentence)
{
     int checkSum;
          int i;
     float convertFactor;

     convertFactor = (360.0/65536.0);
          checkSum = convert2int(&ci_sentence[21]);
          if (checkSum != calcChecksum(ci_sentence, 23))
      {
       return 0;
      }
     else
     {
new_attitude->roll= (float)convert2short(&ci_sentence[1])
                    *convertFactor;
new_attitude->pitch= (float)convert2short(&ci_sentence[3])
```

```
                    *convertFactor;
new_attitude->yaw= (float)convert2short(&ci_sentence[5])
                    *convertFactor;
new_sensor->i_Roll_rate= convert2int(&ci_sentence[13]);
new_sensor->f_Roll_rate= new_sensor->i_Roll_rate/3855.06;
new_sensor->i_Pitch_rate= convert2int(&ci_sentence[15]);
new_sensor->f_Pitch_rate = new_sensor
         ->i_Pitch_rate/3855.06;
        return 1;
    }
}

nodebug root interrupt void input_cap_isr()
{
 capture_status = RdPortI(ICCSR);
  if(capture_status & 0x10)
      {
    //read capture value, LSB first
    PW_Temp = RdPortI(ICL1R)+ RdPortI(ICM1R)*256;
    x++;
    if ((PW_Temp > 4000) || (x>8)) x = 0;
    P_Width[x] = PW_Temp;
    WrPortI(ICCSR, NULL, 0x1C);
      }
}

double strtonum(char* cadena)
{
 double out;
 if(cadena[0]==32)
 cadena[0]=48;

 if(cadena[1]==32)
 cadena[1]=48;

 out= (cadena[0]-48)*100 + (cadena[1]-48)*10 +
      (cadena[2]-48)*1 + (cadena[4]-48)*0.1 +
      (cadena[5]-48)*0.01;

 return out;

}

int sgn (float x)
{
```

```
    if (x>0)
      x=1;
    else
    if (x==0)
      x=0;
    else
      x=-1;
    return x;
}
```

F.2 Backstepping controller in altitude

```
///IMU libraries///
#class auto
#use ''CI3DMGVA4.lib''
///Declaration servo motors outputs///
#define FAST_INTERRUPT 0
#define Servo_Estabilizador 1
#define Servo_Elevador 2
#define Servo_Alabeo 3

///Declaration radio signals///
#define GAS 3
#define YAW 4
#define PITCH 2
#define ROLL 1

///IMU buffer size///
#define BINBUFSIZE 255
#define BOUTBUFSIZE 255

///Xbee buffer size///
#define DINBUFSIZE 255
#define DOUTBUFSIZE 255

///Altimeter buffer size///
#define CINBUFSIZE 7
#define COUTBUFSIZE 7

///MATLAB communication buffer size///
char buff_out[56];

///Cursor position///
nodebug
void gotoxy (int x, int y)
```

```
{
        x += 0x20;
        y += 0x20;
        printf ("\x1B=%c%c",x,y );
}


///Altimeter function declaration///
double strtonum(char*);

///Radio control function declaration///
void input_cap_isr();

///MATLAB functions to send data///
__nodebug
void float2char(float var,int pos)
{
buff_out[pos]  = *((char*)&var);
buff_out[pos+1]= *((char*)&var+1);
buff_out[pos+2]= *((char*)&var+2);
buff_out[pos+3]= *((char*)&var+3);
}

   float yaw,pitch ,roll;
   float yaw_rate,pitch_rate ,roll_rate;
   double z;
   float Pitch_Control;
   float z_des;
   //float superficie;
   float derivada,derivada2;
   double pitch_des;
__nodebug
void enviar_datos()
{
int i,checksum;

checksum=0;
buff_out[0]=0xFF;
buff_out[1]=0x02;
float2char(yaw,2);
float2char(pitch ,6);
float2char(roll ,10);//3
float2char(yaw_rate ,14);
float2char(pitch_rate ,18);
float2char(roll_rate ,22);//6
```

```
float2char(z_des ,26);//7
float2char(z,30);//8
float2char(Pitch_Control ,34);//9
float2char(pitch_des ,38);//10
float2char(derivada ,42);//11
float2char(derivada2 ,46);//12
float2char(0.00,50);

for(i=2;i<54;i++)
checksum += buff_out[i];

buff_out[54]=*((char*)&checksum);
buff_out[55]=*((char*)&checksum+1);

serDwrite(buff_out, sizeof(buff_out));
}

///Variables radio control definitions///
   int PW_Temp , P_Width[8];
   char capture_status;
   int x, manual;

///Main start///
void main()
{
///Variables declaration///
   int i, CI_state, pet, lon;
   unsigned char ID;
   int pwm_options;
   float YawRecta, PitchRecta, RollRecta;
   float Mq,Md;
   unsigned long frecuencia;
   unsigned long t1;
   float td;
   char data[7];
   int Num;
   float e_pitch;
   float k1_p,k2_p,k3_p;
   float Pitch_Final;
   float Control_PitchRecta;
   int Alt_deseado;
   float e_alt;
   //float derivada,derivada2;
   float derivada_anterior,pitch_des_anterior;
```

```
/// Constants declarations ///

    yaw = 0.0;
    pitch = 0.0;
    roll = 0.0;

    yaw_rate = 0.0;
    pitch_rate = 0.0;
    roll_rate = 0.0;

    CI_state = 45;
    pet = 10;
    lon = 45;

    z=0.0;
    pitch_des = 0.0;
          for (i=0; i<45; i++)
          {
                    buffer[i] = 0;
          }

    td=0;
    t1=0;

    Alt_deseado=1;// bandera para amarre de altura deseada
    Pitch_Control=0.0;
    z_des=0.0;
    derivada=0.0;
    derivada2=0.0;
    derivada_anterior=0.0;
    pitch_des_anterior=0.0;

    k1_p = 0.1;
    k2_p = 0.5;
    k3_p = 4;

    Mq = -2.27;
    Md = -2.702;

/// Input capture for radio control ///
#if __SEPARATE_INST_DATA__ && FAST_INTERRUPT
    interrupt_vector inputcap_intvec input_cap_isr;
#else
    SetVectIntern(0x1A, input_cap_isr);
    SetVectIntern(0x1A, GetVectIntern(0x1A));
```

```
#endif
   WrPortI(ICS1R, NULL, 0x44);     // PD1 0x44
   WrPortI(TAT8R, NULL, freq_divider/4-1);
   WrPortI(ICCSR, NULL, 0x1C);
   WrPortI(ICCR, NULL, 0x01);
   WrPortI(ICT1R, NULL, 0x59);

///PWM configuration///
 frecuencia = pwm_init(495);
 pwm_options = PWM_USEPORTE;
 pwm_options |= PWM_OUTEIGHTH;

///Xbee serial port configuration///
   serDopen(115200);
   serDwrFlush();
   serDrdFlush();

///IMU serial port configuration///
        serBopen(115200);
        creaDivision(division, 10, 0);
        serBwrite(division, 9);
        creaPeticion(peticion, pet);

///Altimeter serial port configuration///
    serCopen(38400);
        serCwrFlush();
        serCrdFlush();

///Central reading cycle start and manual mode///
 while(1)
        {
   costate
              {
                    if (CI_state == lon)
                       {
                          wfd  cof_serBwrite(peticion, 6);
                          CI_state = 0;
                       }
                        CI_state = serBread(buffer, lon, 5);
                        yield;
               }

  costate
  {
 if (buffer[0]== 0x81 &&  buffer[1]== 0xA1
```

```
                        && buffer[2]== 0x0A){
                          StoreNAV_SENSOR(&cent_ins , buffer);
                  yaw =   0.01*(cent_ins.ypr[0]);
                  pitch = 0.01*(cent_ins.ypr[1]);
                  roll =  0.01*(cent_ins.ypr[2]);
                  yaw_rate =    0.01*(cent_ins.pqr[2]);
                  pitch_rate =  0.01*(cent_ins.pqr[1]);
                  roll_rate =   0.01*(cent_ins.pqr[0]);
              }
            }

    costate {
                  serCrdFlush();
                   wfd  Num = cof_serCread(data,7,100);
                   if(data[6]==0x0B)
                  {
                     z=strtonum(data);

                  }
                      }

///Servo motors control///
costate{
              YawRecta = 0.4039 * P_Width[YAW] + 210.9634;
              if (YawRecta < 566 )
                 {
                   YawRecta = 566;
                 }
              if (YawRecta > 982)
                 {
                   YawRecta = 982;
                 }
     pwm_set(Servo_Estabilizador , (int)(YawRecta),
             pwm_options);
          }

costate{
   if (P_Width[5] < 1000)
        {
            PitchRecta = 0.4039 * P_Width[PITCH] + 210.9634;
             if (PitchRecta < 566)
                 {
                   PitchRecta = 566;
                 }
             if (PitchRecta > 982)
```

```
                {
                  PitchRecta = 982;
                }
pwm_set(Servo_Elevador, (int)(PitchRecta), pwm_options);
            Alt_deseado = 1;
            Pitch_Control = 0.0;
            derivada = 0.0;
            derivada2 = 0.0;
        }
      }

///Backstepping control in altitude definition///
costate{
if(P_Width[5]>1000&&buffer[0]==0x81&&buffer[1]
                    ==0xA1&&buffer[2]== 0x0A)
        {
        waitfor(DelayMs(1));
        if(Alt_deseado==1)
          {
          z_des = z;
          Alt_deseado=0;
          }
          pitch_des= deg(atan2(z_des-z,14));
          e_pitch = pitch_des - pitch;
          td = (MS_TIMER - t1)/1000.0;
          t1 = MS_TIMER;

  derivada=sat((pitch_des - pitch_des_anterior)/td,
          -5.0, 5.0);
          pitch_des_anterior = pitch_des;

  derivada2 = sat((derivada - derivada_anterior)/td,
            -5.0, 5.0);
          derivada_anterior = derivada;

 Pitch_Control = ((Mq*pitch_rate - k1_p
                 *(derivada-pitch_rate)
                 - derivada2 - k3_p*e_pitch
                 - k2_p*(derivada + k1_p*e_pitch
                 - pitch_rate))
                 /(-Md))-15.696;

          Control_PitchRecta = 0.4039 *  Pitch_Control
                                   + 774;
           if (Control_PitchRecta < 566)
```

```
                {
                  Control_PitchRecta = 566;
                }
              if (Control_PitchRecta > 982)
                {
                  Control_PitchRecta = 982;
                }

pwm_set(Servo_Elevador, (int)(Control_PitchRecta),
        pwm_options);
      }
    }

costate{
              RollRecta = 0.4000 * P_Width[ROLL] + 180;
              if (RollRecta < 566)
                {
                  RollRecta = 566;
                }
              if (RollRecta > 982)
                {
                  RollRecta = 982;
                }
              pwm_set(Servo_Alabeo, (int)(RollRecta),
                      pwm_options);
                }

 costate{
                  serDwrFlush();
                              serDrdFlush();
                  enviar_datos();
                  waitfor(DelayMs(25));
 }
        }
}

///Cofunction definitions///

unsigned short creaPeticion(unsigned char *buffer,
                            unsigned char ID)
{
        unsigned char *p, *p_end;
        unsigned char pay_len;
        unsigned char ck0;
        unsigned char ck1;
```

```
        ck0 = 0;
        ck1 = 0;
        buffer[0] = 0x81;
        buffer[1] = 0xA1;
        buffer[2] = ID;
        buffer[3] = 0x00;
        p_end = &buffer[4];
        pay_len = p_end - buffer - 4;
        buffer[3] = pay_len;
        p=buffer+2;
        while(p < p_end)
        {
                ck0 += *p++;
                ck1 += ck0;
        }
        *p++ = ck0;
        *p++ = ck1;

        return(p-buffer);
}

unsigned short creaDivision(unsigned char *buffer,
                            unsigned char mens,
                            unsigned char div)
{
        unsigned char *p, *p_end;
        unsigned char pay_len;
        unsigned char ck0;
        unsigned char ck1;
        ck0 = 0;
        ck1 = 0;
        buffer[0] = 0x81;
        buffer[1] = 0xA1;
        buffer[2] = 0x23;
        buffer[3] = 0x03;
        buffer[4] = 0x05;
        buffer[5] = mens;
        buffer[6] = div;
        p_end = &buffer[7];
        pay_len = p_end - buffer - 4;
        buffer[3] = pay_len;
        p=buffer+2;
        while(p < p_end)
        {
                ck0 += *p++;
```

```
                ck1 += ck0;
        }

        *p++ = ck0;
        *p++ = ck1;
        return(p-buffer);
}

nodebug static void StoreNAV_SENSOR(mtMIDG2_NAV_SENSOR *p
                                    , unsigned char *buf)
  {
  int i;
        ++buf;
        ++buf;
        ++buf;
    ++buf;

  p->Time=mBinGetULong(&buf);
  for(i=0;i<3;++i)
    p->pqr[i]=mBinGetShort(&buf);
  for(i=0;i<3;++i)
    p->axyz[i]=mBinGetShort(&buf);
  for(i=0;i<3;++i)
    p->ypr[i]=mBinGetShort(&buf);
  for(i=0;i<4;++i)
    p->Q[i]=mBinGetLong(&buf);
    p->Flags=*buf++;
    p->updated=1;
  }

nodebug root interrupt void input_cap_isr()
{
 capture_status = RdPortI(ICCSR);
 if(capture_status & 0x10)
      {
         PW_Temp = RdPortI(ICL1R)+ RdPortI(ICM1R)*256;
         x++;
         if ((PW_Temp > 4000) || (x>8)) x = 0;
         //cambio 6-->8
         P_Width[x] = PW_Temp;
         WrPortI(ICCSR, NULL, 0x1C); //zero out counter1
      }
}

double strtonum(char* cadena)
```

```
{
 double out;
 if(cadena[0]==32)
 cadena[0]=48;

 if(cadena[1]==32)
 cadena[1]=48;

 out= (cadena[0]-48)*100 + (cadena[1]-48)*10
        + (cadena[2]-48)*1
        + (cadena[4]-48)*0.1 + (cadena[5]-48)*0.01;

 return out;
}
```

Linear and nonlinear state observers: programs for the embedded system

G

In Section G.1 the program used in the real flight test with the Luenberger observer with PD controller for yaw is presented, and in Section G.2 the program for SMO with PD controller for altitude is presented.

G.1 Luenberger observer with PD controller in yaw

```
///IMU libraries///
#class auto
#use "CI3DMGVA4.lib"

///Declaration servo motors outputs///
#class auto
#define FAST_INTERRUPT 0
#define Servo_Estabilizador 1
#define Servo_Elevador 2
#define Servo_Alabeo  3

///Declarations radio signals///
#define GAS   3
#define YAW   4
#define PITCH 2
#define ROLL  1

///IMU buffer size///
#define BINBUFSIZE  31
#define BOUTBUFSIZE 31

///Xbee buffer size///
#define DINBUFSIZE  255
#define DOUTBUFSIZE 255

///Altimeter buffer size///
#define CINBUFSIZE  7
#define COUTBUFSIZE 7
```

```
///MATLAB communication buffer size///
char buff_out[56];

///Structure to read the data from IMU///
typedef struct {
   int i_Roll_rate;
   int i_Pitch_rate;
   int i_Yaw_rate;
   unsigned int u_Roll_rate;
   unsigned int u_Pitch_rate;
   unsigned int u_Yaw_rate;
   float f_Roll_rate;
   float f_Pitch_rate;
   float f_Yaw_rate;
   float Acc_X;
   float Acc_Y;
   float Acc_Z;
   int Mag_X;
   int Mag_Y;
   int Mag_Z;
}Sensores_CI;

///Structure to read the data from IMU///
typedef struct {
   float pitch;
   float roll;
   float yaw;
   } Attitude;

///Variables declaration///
float PitchRateGrad,RollRateGrad,YawRateGrad;
double z;
float yaw_des;
float Yaw_Control;
float Psi_ei,Psi2_ei;
float T_obs;
float Psi_ep,Psi2_ep;
float K1,K2;
float Nr,Nd;



///IMU function///
int obtiene_angulo(Attitude *new_attitude,
```

```
                    Sensores_CI *new_sensor,
                    char *ci_sentence);

///Function to read radio control signals///
void input_cap_isr();

///Sign function///
float sgn (float x);

///Cursor position///
nodebug void gotoxy (int x, int y)
{
   x += 0x20;
   y += 0x20;
   printf ("\x1B=%c%c",x,y );
}

///IMU variables///
Attitude  Actual_central;
Attitude Actual_Attitude;
Sensores_CI Actual_sensor;
char Buffer_Central[73];
int CI_state;

///Manual control variables///
int  PW_Temp , P_Width[8];
char capture_status;
int  x, i, manual;

/// Altimeter function///
double strtonum(char*);
double z;

///MATLAB functions to send data///
__nodebug
void float2char(float var,int pos)
{
buff_out[pos]  = *((char*)&var);
buff_out[pos+1]= *((char*)&var+1);
buff_out[pos+2]= *((char*)&var+2);
buff_out[pos+3]= *((char*)&var+3);
}
__nodebug
void enviar_datos()
{
```

```
int i,checksum;

checksum=0;
buff_out[0]=0xFF;
buff_out[1]=0x02;
float2char(Actual_Attitude.roll ,2);
float2char(Actual_Attitude.pitch ,6);
float2char(Actual_Attitude.yaw,10);
float2char(RollRateGrad ,14);
float2char(PitchRateGrad ,18);
float2char(YawRateGrad ,22);//6
float2char(yaw_des ,26);
float2char(Yaw_Control ,30);
float2char(Psi_ei ,34);
float2char(Psi2_ei ,38);
float2char(T_obs ,42);
float2char(Psi2_ep ,46);
float2char(0.00 ,50);

for(i=2;i<54;i++)
checksum += buff_out[i];

buff_out[54]=*((char*)&checksum);
buff_out[55]=*((char*)&checksum+1);

serDwrite(buff_out , sizeof(buff_out));
}

///Main start///
void main()
{
///Variables declaration///

    int   roll , pitch , yaw;
    int   pwm_options;
    float YawRecta, PitchRecta , RollRecta;
    double T;
    unsigned long T1,T1_obs;
    unsigned long  frecuencia;
    char data[7];
    int Num;
    float ep;
    float kp_y,kv_y;
    float Yaw_Final;
    float Control_YawRecta;
```

```
    int Yaw_deseado;

/// Constants declarations ///

    roll = 0;
    pitch = 0;
    yaw = 0;

    Actual_central.roll = 0.0;
    Actual_central.pitch = 0.0;
    Actual_Attitude.yaw = 0.0;

    Actual_sensor.f_Roll_rate = 0.0;
    Actual_sensor.f_Pitch_rate = 0.0;
    Actual_sensor.f_Yaw_rate = 0.0;

    YawRecta = 0.0;
    PitchRecta = 0.0;
    RollRecta = 0.0;

    RollRateGrad=0.0;
    YawRateGrad=0.0;
    PitchRateGrad=0.0;

    T = 0;
    T1 = 0;
    T_obs=0.0;
    T1_obs=0;

    Yaw_deseado=1;
    Yaw_Control=0.0;
    yaw_des=0.0;

    kp_y = 5.0;//0.8
    kv_y = 1.0;//8.0

    Psi_ep=0.0;
    Psi2_ep=0.0;
    Psi_ei=0.0;
    Psi2_ei=0.0;

    Nr = -16.4469*0.01;
    Nd = -599.8125*0.002;

    K1 = 10.0;
```

```
    K2 = 180.0;

///Input capture for radio control///
#if __SEPARATE_INST_DATA__ && FAST_INTERRUPT
    interrupt_vector inputcap_intvec input_cap_isr;
#else
    SetVectIntern(0x1A, input_cap_isr);
    SetVectIntern(0x1A, GetVectIntern(0x1A));
#endif

    P_Width[0] = 0;

    WrPortI(ICS1R, NULL, 0x44);
    WrPortI(TAT8R, NULL, freq_divider/4-1);
    WrPortI(ICCSR, NULL, 0x1C);
    WrPortI(ICCR, NULL, 0x01);
    WrPortI(ICT1R, NULL, 0x59);

///PWM configuration///
  frecuencia = pwm_init(495);
  pwm_options = PWM_USEPORTE;
  pwm_options |= PWM_OUTEIGHTH;

///IMU serial configuration///
    serBopen(115200);
    serBputc(0x06);
    serBread(Buffer_Central,5,5);
    serBrdFlush();
    CI_state = 0;

///Xbee serial port configuration///
    serDopen(115200);
    serDwrFlush();
    serDrdFlush();
    T1=MS_TIMER;

///Altimeter serial port configuration///
    serCopen(38400);
    serCwrFlush();
    serCrdFlush();

///Central reading cycle start and manual mode///
 while(1)
    {
costate
```

```
{
 wfd   cof_serBputc(0x31);
 wfd   CI_state = cof_serBread(Buffer_Central,23,5);
 yield;
 if  (CI_state == 23 && Buffer_Central[0]== 0x31)
{
obtiene_angulo(&Actual_Attitude,&Actual_sensor,
                   Buffer_Central);
CI_state = 0;
Buffer_Central[0]=0x00;

///Luenberger observer definition///
Psi_ep = Psi2_ei + K1*(Actual_Attitude.yaw - Psi_ei);
Psi2_ep = Nr*Psi2_ei + Nd*Yaw_Control
          + K2*(Actual_Attitude.yaw - Psi_ei) ;

   T_obs = ((double) (MS_TIMER - T1_obs))/1000.0;
   T1_obs = MS_TIMER;
   Psi_ei = Psi_ei+Psi_ep*T_obs;
   Psi2_ei = Psi2_ei+Psi2_ep*T_obs;
                    }
              }

  costate {
                serCrdFlush();
                 wfd  Num = cof_serCread(data,7,100);
                 if(data[6]==0x0B)
                {
                   z=strtonum(data);
                 }
                }

  costate{
                RollRateGrad=Actual_sensor.f_Roll_rate
                              *57.2958;
                YawRateGrad=Actual_sensor.f_Yaw_rate
                             *57.2958;
                PitchRateGrad=Actual_sensor.f_Pitch_rate
                              *57.2958;
          }

///Servo motors control///
costate{
if (P_Width[5]<1000&&CI_state == 23
                    && Buffer_Central[0]== 0x31)
```

```
        {
            YawRecta = 0.4000 * P_Width[YAW] + 180;
            if (YawRecta < 546 )
                {
                  YawRecta = 546;
                }
            if (YawRecta > 946)
                {
                  YawRecta = 946;
                }

pwm_set(Servo_Estabilizador, (int)(YawRecta), pwm_options);
           Yaw_deseado=1;
           Yaw_Control = 0.0;
        }
      }

costate{

if(P_Width[5]>1000&&CI_state == 23
                    && Buffer_Central[0]== 0x31)
        {

           if(Yaw_deseado==1)
           {
           yaw_des= Actual_Attitude.yaw;
           Yaw_deseado=0;
           }
           ep = Psi_ei - yaw_des;
           Yaw_Control = -kp_y*ep - kv_y*Psi2_ei;
           Control_YawRecta = 1 * Yaw_Control + 746;
            if (Control_YawRecta < 546 )
                {
                  Control_YawRecta = 546;
                }
            if (Control_YawRecta > 946)
                {
                  Control_YawRecta = 946;
                }
pwm_set(Servo_Estabilizador,(int)(Control_YawRecta),
          pwm_options);

          }
        }
```

```
costate{
  PitchRecta = 0.4000 * P_Width[PITCH] + 180;
            if (PitchRecta < 546)
               {
                 PitchRecta = 546;
               }
            if (PitchRecta > 946)
               {
                 PitchRecta = 946;
               }
pwm_set(Servo_Elevador, (int)(PitchRecta), pwm_options);
           }

costate{
            RollRecta = 0.4000 * P_Width[ROLL] + 180;
            if (RollRecta < 546)
               {
                 RollRecta = 546;
               }
            if (RollRecta > 946)
               {
                 RollRecta = 946;
               }

pwm_set(Servo_Alabeo, (int)(RollRecta), pwm_options);
          }
costate{
                 serDwrFlush();
                             serDrdFlush();
                 enviar_datos();
                 waitfor(DelayMs(100));

         }
    }
}


///Cofunctions definition///

nodebug int obtiene_angulo(Attitude *new_attitude,
        Sensores_CI *new_sensor,  char *ci_sentence)
{
    int checkSum;
    int i;
    float convertFactor;
```

```
    convertFactor = (360.0/65536.0);
    checkSum = convert2int(&ci_sentence[21]);
    if (checkSum != calcChecksum(ci_sentence , 23))
     {
      return 0;
     }
    else
    {
new_attitude->roll= (float)convert2short(&ci_sentence[1])
                    *convertFactor;
new_attitude->pitch= (float)convert2short(&ci_sentence[3])
                    *convertFactor;
new_attitude->yaw= (float)convert2short(&ci_sentence[5])
                    *convertFactor;
new_sensor->i_Roll_rate= convert2int(&ci_sentence[13]);
new_sensor->f_Roll_rate= new_sensor->i_Roll_rate/3855.06;
new_sensor->i_Pitch_rate= convert2int(&ci_sentence[15]);
new_sensor->f_Pitch_rate = new_sensor
         ->i_Pitch_rate/3855.06;
new_sensor->i_Yaw_rate = convert2int(&ci_sentence[17]);
new_sensor->f_Yaw_rate = new_sensor->i_Yaw_rate/3855.06;

        return 1;
    }

}

nodebug root interrupt void input_cap_isr()
{
 capture_status = RdPortI(ICCSR);
 if(capture_status & 0x10)
     {
        PW_Temp = RdPortI(ICL1R)+ RdPortI(ICM1R)*256;
        x++;
        if ((PW_Temp > 4000) || (x>8)) x = 0;
        P_Width[x] = PW_Temp;
        WrPortI(ICCSR, NULL, 0x1C);
     }
}

double strtonum(char* cadena)
{
 double out;
 if(cadena[0]==32)
 cadena[0]=48;
```

```
  if(cadena[1]==32)
  cadena[1]=48;

  out= (cadena[0]-48)*100 + (cadena[1]-48)*10
      + (cadena[2]-48)*1 + (cadena[4]-48)*0.1
      + (cadena[5]-48)*0.01;

  return out;
}

float sgn (float x)
{
float s;
  if (x>0)
    s = 1.0;
  else
  if (x==0)
    s = 0.0;
  else
    s = -1.0;
  return s;

  //if (x<0)
}
```

G.2 SMO observer with PD controller in altitude

```
///IMU libraries///
#class auto
#use "CI3DMGVA4.lib"

///Declaration servo motors outputs///
#class auto
#define FAST_INTERRUPT 0
#define Servo_Estabilizador 1
#define Servo_Elevador 2
#define Servo_Alabeo  3

///Declaration radio signals///
#define GAS   3
#define YAW   4
#define PITCH 2
#define ROLL  1
```

```
///IMU buffer size///
#define BINBUFSIZE  31
#define BOUTBUFSIZE 31

///Xbee bugger size///
#define DINBUFSIZE  255//63
#define DOUTBUFSIZE 255//63

/// Altimeter buffer size///
#define CINBUFSIZE  7
#define COUTBUFSIZE 7

///MATLAB communication buffer size///
char buff_out[56];


/// Structure to read the data from IMU///
typedef struct {
        int i_Roll_rate;
        int i_Pitch_rate;
        int i_Yaw_rate;
        unsigned int u_Roll_rate;
        unsigned int u_Pitch_rate;
        unsigned int u_Yaw_rate;
    float f_Roll_rate;
    float f_Pitch_rate;
    float f_Yaw_rate;
    float Acc_X;
    float Acc_Y;
    float Acc_Z;
    int Mag_X;
    int Mag_Y;
    int Mag_Z;
}Sensores_CI;

/// Structure to read the data from IMU///
typedef struct {
        float pitch;
        float roll;
        float yaw;
        } Attitude;

    float PitchRateGrad,RollRateGrad,YawRateGrad;
    double z;
    double z_des;
```

```
    float Pitch_Control;
    float Psi_ei, Psi2_ei;

///Instruction to read IMU///
int obtiene_angulo(Attitude *new_attitude,
 Sensores_CI *new_sensor, char *ci_sentence);

///Manual control///
void input_cap_isr();

///Sign function///
float sgn (float x);

///Cursor position///
nodebug void gotoxy (int x, int y)
{
        x += 0x20;
        y += 0x20;
        printf ("\x1B=%c%c",x,y );
}

///IMU Variables///
Attitude  Actual_central;
Attitude Actual_Attitude;
Sensores_CI Actual_sensor;
char Buffer_Central[23];
int CI_state;

///Manual control variables///
int  PW_Temp , P_Width[8];
char capture_status;
int  x, i, manual;

/// Altimeter functions///
double strtonum(char*);
double z;

///MATLAB functions to send data///
__nodebug
void float2char(float var,int pos)
{
buff_out[pos]  = *((char*)&var);
buff_out[pos+1]= *((char*)&var+1);
buff_out[pos+2]= *((char*)&var+2);
buff_out[pos+3]= *((char*)&var+3);
```

```
}

__nodebug
void enviar_datos()
{
int i,checksum;

checksum=0;
buff_out[0]=0xFF;
buff_out[1]=0x02;
float2char(Actual_Attitude.roll ,2);//1
float2char(Actual_Attitude.pitch ,6);//2
float2char(Actual_Attitude.yaw,10);//3
float2char(RollRateGrad ,14);//4
float2char(PitchRateGrad ,18);//5
float2char(YawRateGrad,22);//6
float2char(z_des ,26);//7
float2char(Pitch_Control ,30);//8
float2char(z,34);//9
float2char(Psi_ei ,38);//10
float2char(Psi2_ei ,42);//11
float2char(0.00,46);
float2char(0.00,50);

for(i=2;i<54;i++)
checksum += buff_out[i];

buff_out[54]=*((char*)&checksum);
buff_out[55]=*((char*)&checksum+1);

serDputc(buff_out[i]);
}

///Main start///
void main()
{
///Variables declaration///
   int  roll, pitch, yaw;
   int   pwm_options;
   float YawRecta, PitchRecta, RollRecta;
   double T,T_obs,t;
   unsigned long T1,T1_obs,t1;
   unsigned long  frecuencia;
   char data[7];
   int Num;
```

```
    float e_pitch;
    float pitch_des;
    float kp_p,kv_p;
    float Pitch_Final;
    float Control_PitchRecta;
    int Alt_deseado;
    float e_alt;
    float derivada;
    float e_pitch_anterior;
    float Psi_ep,Psi2_ep;
    float alfa1,alfa2,gama1,gama2;

///Constants declarations///

    roll = 0;
    pitch = 0;
    yaw = 0;

    Actual_central.roll = 0.0;
    Actual_central.pitch = 0.0;
    Actual_central.yaw = 0.0;

    Actual_sensor.f_Roll_rate = 0.0;
    Actual_sensor.f_Pitch_rate = 0.0;
    Actual_sensor.f_Yaw_rate = 0.0;

    YawRecta = 0.0;
    PitchRecta = 0.0;
    RollRecta = 0.0;

    RollRateGrad=0.0;
    YawRateGrad=0.0;
    PitchRateGrad=0.0;

    T = 0;
    T1 = 0;
    t=0;
    t1=0;


    Alt_deseado=1;// bandera para amarre de altura deseada
    Pitch_Control=0.0;
    z_des=0.0;

    kp_p = 10;//0.8 ,sim=2 ,6.0,10
```

```
    kv_p = 0.05;//8.0 ,sim=4 ,8.0,0.1

    Psi_ep=0.0;
    Psi2_ep=0.0;
    Psi_ei=0.0;
    Psi2_ei=0.0;

    alfa1=55;
    alfa2=570;
    gama1=0.1;
    gama2=0.85;

///Input capture for radio control///

#if __SEPARATE_INST_DATA__ && FAST_INTERRUPT
    interrupt_vector inputcap_intvec input_cap_isr;
#else
    SetVectIntern(0x1A, input_cap_isr);
    SetVectIntern(0x1A, GetVectIntern(0x1A));
#endif

    P_Width[0] = 0;

    WrPortI(ICS1R, NULL, 0x44);
    WrPortI(TAT8R, NULL, freq_divider/4-1);
    WrPortI(ICCSR, NULL, 0x1C);
    WrPortI(ICCR, NULL, 0x01);
    WrPortI(ICT1R, NULL, 0x59);

///PWM configuration///
  frecuencia = pwm_init(495);
  pwm_options = PWM_USEPORTE;
  pwm_options |= PWM_OUTEIGHTH;

///IMU serial port configuration///
    serBopen(115200);
    serBputc(0x06);
    serBread(Buffer_Central,5,5);//valor2=0
    serBrdFlush();
    CI_state = 0;

///Xbee serial port configuration///
    serDopen(115200);
    serDwrFlush();
    serDrdFlush();
```

```
    T1=MS_TIMER;

/// Altimeter serial port configuration///
    serCopen(38400);
    serCwrFlush();
    serCrdFlush();

/// Central reading cycle start and manual mode///
  while(1)
{
costate
{
wfd  cof_serBputc(0x31);
wfd  CI_state = cof_serBread(Buffer_Central ,23 ,5);
yield;
if (CI_state == 23 && Buffer_Central[0]== 0x31)
{
obtiene_angulo(&Actual_Attitude ,&Actual_sensor ,
                 Buffer_Central);
CI_state = 0;
Buffer_Central[0]=0x00;

///SMO observer definition///
Psi_ep = -alfa1*(Psi_ei-Actual_Attitude.pitch)
+ Psi2_ei - gama1*sgn(Psi_ei-Actual_Attitude.pitch);
Psi2_ep = -alfa2*(Psi_ei-Actual_Attitude.pitch)
- gama2*sgn(Psi_ei-Actual_Attitude.pitch) - 0.0128*Psi2_ei
- 2.3625*0.01*Pitch_Control ;

                  T_obs = ((double) (MS_TIMER - T1_obs))
                         /1000.0;
                  T1_obs = MS_TIMER;
                  Psi_ei = Psi_ei+Psi_ep*T_obs;
                  Psi2_ei = Psi2_ei+Psi2_ep*T_obs;
                                  }
                         }

  costate {
                 serCrdFlush();
                 wfd  Num = cof_serCread(data ,7 ,100);
                 if(data[6]==0x0B)
               {
                   z=strtonum(data);
                 }
                    }
```

```
   costate{
                RollRateGrad=Actual_sensor.f_Roll_rate
                                *57.2958;
                YawRateGrad=Actual_sensor.f_Yaw_rate
                                *57.2958;
                PitchRateGrad=Actual_sensor.f_Pitch_rate
                                *57.2958;
         }


///Servo motors control///
costate{
           YawRecta = 0.4000 * P_Width[YAW] + 180;
           if (YawRecta < 546 )
              {
                YawRecta = 546;
              }
           if (YawRecta > 946)
              {
                YawRecta = 946;
              }

pwm_set(Servo_Estabilizador, (int)(YawRecta),
         pwm_options);
       }

costate{
if(P_Width[5]<1000&& CI_state == 23
                     && Buffer_Central[0]== 0x31)
       {
           PitchRecta = 0.4000 * P_Width[PITCH] + 180;
           if (PitchRecta < 546)
              {
                PitchRecta = 546;
              }
           if (PitchRecta > 946)
              {
                PitchRecta = 946;
              }

pwm_set(Servo_Elevador, (int)(PitchRecta), pwm_options);
           Alt_deseado = 1;
           Pitch_Control = 0.0;
       }
```

```
        }
costate{
if(P_Width[5]>1000&& CI_state == 23
                    && Buffer_Central[0]== 0x31)
        {
        if(Alt_deseado==1)
          {
          z_des = z;
          Alt_deseado=0;
          }
           e_alt=z_des-z;
           pitch_des= atan2(z_des-z,14);
           e_pitch= - Psi_ei + deg(pitch_des);

          t = ((double) (MS_TIMER - t1))/1000.0;
          t1 = MS_TIMER;
          derivada = (e_pitch - e_pitch_anterior)/(float)t;
          e_pitch_anterior = e_pitch;

          Pitch_Control = -kp_p*e_pitch - kv_p*derivada;

          Control_PitchRecta = 1 *  Pitch_Control + 746;
           if (Control_PitchRecta < 546)
              {
                Control_PitchRecta = 546;
              }
           if (Control_PitchRecta > 946)
              {
                Control_PitchRecta = 946;
              }

          pwm_set(Servo_Elevador, (int)(Control_PitchRecta)
                  , pwm_options);
          }
       }

costate{
           RollRecta = 0.4000 * P_Width[ROLL] + 180;
           if (RollRecta < 546)
              {
                RollRecta = 546;
              }
           if (RollRecta > 946)
              {
                RollRecta = 946;
```

```
                }

pwm_set(Servo_Alabeo, (int)(RollRecta), pwm_options);
}

costate{
        serDwrFlush();
        serDrdFlush();
        enviar_datos();
        waitfor(DelayMs(50));// antes 25
       }
        }
}

///Cofunctions definition///

nodebug int obtiene_angulo(Attitude *new_attitude,
Sensores_CI *new_sensor, char *ci_sentence)
{
    int checkSum;
         int i;
    float convertFactor;

    convertFactor = (360.0/65536.0);
         checkSum = convert2int(&ci_sentence[21]);
         if (checkSum != calcChecksum(ci_sentence, 23))
     {
      return 0;
     }
    else
    {
new_attitude->roll= (float)convert2short(&ci_sentence[1])
                    *convertFactor;
new_attitude->pitch= (float)convert2short(&ci_sentence[3])
                    *convertFactor;
new_attitude->yaw= (float)convert2short(&ci_sentence[5])
                    *convertFactor;
new_sensor->i_Roll_rate = convert2int(&ci_sentence[13]);
new_sensor->f_Roll_rate = new_sensor->i_Roll_rate/3855.06;
new_sensor->i_Pitch_rate = convert2int(&ci_sentence[15]);
new_sensor->f_Pitch_rate = new_sensor
          ->i_Pitch_rate/3855.06;
new_sensor->i_Yaw_rate = convert2int(&ci_sentence[17]);
new_sensor->f_Yaw_rate = new_sensor->i_Yaw_rate/3855.06;
        return 1;
```

```
    }
}

nodebug root interrupt void input_cap_isr()
{
capture_status = RdPortI(ICCSR);
//listen for capture states
if(capture_status & 0x10)
// channel 1 stop condition detected
      {
          //read capture value, LSB first
          PW_Temp = RdPortI(ICL1R)+ RdPortI(ICM1R)*256;
          x++;
          if ((PW_Temp > 4000) || (x>8)) x = 0;
          ,//cambio 6-->8
          P_Width[x] = PW_Temp;
          WrPortI(ICCSR, NULL, 0x1C); //zero out counter1
      }
}

double strtonum(char* cadena)
{
 double out;
 if(cadena[0]==32)
 cadena[0]=48;

 if(cadena[1]==32)
 cadena[1]=48;

 out= (cadena[0]-48)*100 + (cadena[1]-48)*10
     + (cadena[2]-48)*1 + (cadena[4]-48)*0.1
     + (cadena[5]-48)*0.01;

 return out;
}

float sgn (float x)
{
float s;
  if (x>0)
    s = 1.0;
  else
  if (x==0)
    s = 0.0;
  else
```

```
        s = -1.0;
    return s;

    //if (x<0)
}
```

MATLAB® program to graph

H

This section presents the MATLAB® program to graph the results obtained from the fixed-wing UAV in the real flight tests.

```
clear all;
close all;
clc;
tic

e=0;
buffer_in = zeros(1,13);

PS=serial('COM10');
scopy = PS;
clear PS
PS = instrfind;
set(PS,'Baudrate',115200);
set(PS,'StopBits',1);
set(PS,'DataBits',8);
set(PS,'Parity','none');
set(PS,'Timeout',2);
fopen(PS);

count = 0;

fileID = fopen('File.txt','a');
disp('Start...');

while 1
    %pause(2);

    [request,count] = fread(PS,56,'uchar');% uint8
    if (count == 56 && request(1) == 255
                    && request(2) == 2)
    buffer_in = checksum(request);
    for i=1:13
            fprintf('%10.4f\t',buffer_in(i));
    end
        fprintf('\n\r');

        fprintf(fileID,'\t%10.4f\t%10.4f',buffer_in,toc);
```

```
        fprintf(fileID ,'\n');
    else
    disp('Bad data!!');
    e=e+1;
    fclose(PS);
    fopen(PS);
    end
end
```

Altimeter program

This appendix presents the program to read the altitude sensor from the Parallax microcontroller after sending the data to be read from the Rabbit microprocessor.

```
OBJ

  pst    : "parallax serial terminal"
  alt    : "29124_altimeter"

CON

  _clkmode        = xtal1 + pll16x
  _xinfreq        = 5_000_000

'Your starting altitude in feet.
 You can get this from a local airport 1317
  START_ALT       = 10

PUB start | a, p, t

  pst.start(38400)' Start Parallax serial terminal.
  alt.start(alt#QUICKSTART, alt#BACKGROUND)'
                                          Start altimeter.
  alt.set_resolution(alt#HIGHEST) '
                              Set to highest resolution.
  alt.set_altitude(alt.m_from_ft(START_ALT * 100))'
                                        Starting altitude.
    a := alt.altitude(alt.average_press)'
                                 Get the current altitude
     HM = home
    pst.str(alt.formatn(a, alt#CECR, 5))'
                                 Print altitude in meters

    if (pst.rxcount) '
                   Respond to any key by clearing screen.
      pst.rxflush
      pst.char(pst#CS)
```

Bibliography

[1] T.M. Adami, J.J. Zhu, 6 DOF flight control of fixed-wing aircraft by trajectory linearization, in: American Control Conference (ACC), San Francisco, California, USA, June–July, 2011.

[2] B.R. Andrievsky, A.L. Fradkov, Combined adaptive controller for UAV guidance, European Journal of Control 11 (1) (2005) 71–79.

[3] K.J. Åström, B. Wittenmark, Adaptive Control, 2nd edition, Prentice Hall, USA, 1994.

[4] N. Bouton, R. Lenain, B. Thuilot, F. Martinet, A tire stiffness backstepping observer dedicated to all-terrain vehicle rollover prevention, Advanced Robotics 22 (2008) 1267–1285.

[5] M. Boutayeb, M. Darouach, Comments on a robust state observer scheme, IEEE Transactions on Automatic Control 48 (7) (Jul. 2003) 1292–1294.

[6] A. Brezoescu, T. Espinoza, P. Castillo, R. Lozano, Adaptive trajectory following for a fixed-wing UAV in presence of crosswind, Journal of Intelligent & Robotic Systems 69 (1–4) (2013) 257–271.

[7] R. Beard, D. Kingston, M. Quigley, D. Snyder, R. Christiansen, Autonomous vehicle technologies for small fixed-wing UAVs, Journal of Aerospace Computing, Information, and Communication 2 (1) (2005) 92–108.

[8] D.B. Barber, J.D. Redding, T.W. Mclain, R.W. Beard, C.N. Taylor, Vision-based target geo-location using a fixed-wing miniature air vehicle, Journal of Intelligent & Robotic Systems 47 (4) (2006) 361–382.

[9] R. Beard, T. Mclain, Small Unmanned Aircraft Theory and Practice, Princeton University Press, United Kingdom, 2012.

[10] G. Bartolini, L. Fridman, A. Pisano, E. Usai, Modern Sliding Mode Control Theory, Springer, USA, 2008.

[11] Y.H. Chen, Adaptive robust observers for non-linear uncertain systems, International Journal of Systems Science 21 (1990) 803–814.

[12] C.T. Chen, Linear Systems Theory and Design, Oxford University Press, Inc., September 1998.

[13] J.J. Craig, Introduction to Robotics Mechanics and Control, third edition, Pearson Education, USA, 2005.

[14] M.V. Cook, Flight Dynamics, second edition, Betterworth-Heinemann, USA, 2007.

[15] H. Castañeda, O.S. Peña, J.L. Morales, Adaptive super twisting flight control-observer for a fixed wing UAV, in: International Conference on Unmanned Aircraft Systems (ICUAS), Atlanta, GA, USA, May 28–31, 2013.

[16] Y.H. Chen, Adaptive robust observers for non-linear uncertain systems, International Journal of Systems Science 21 (1990) 803–814.

[17] J. Deleon, I. Souleiman, A. Glumineau, G. Schireier, On nonlinear equivalence and backstepping observer, Kybernetika 37 (5) (2001) 521–546.

[18] L. Dickerson, UAVs on the rise, Aviation Week & Space Technology, Aerospace Source Book 2007 166 (3) (2007).

[19] D.M. Dawson, Z. Qu, J.C. Carroll, On the state observation and output feedback problems for nonlinear uncertain dynamic systems, Systems & Control Letters 18 (1992) 217–222.

[20] M. Darouach, Linear functional observers for systems with delays in state variables, IEEE Transactions on Automatic Control 46 (5) (Mar. 2001) 491–496.
[21] M. Darouach, Linear functional observers for systems with delays in state variables: the discrete-time case, IEEE Transactions on Automatic Control 50 (2) (Feb. 2005) 228–233.
[22] M. Darouach, Reduced-order observers for linear neutral delay systems, IEEE Transactions on Automatic Control 50 (9) (Sep. 2005) 1407–1413.
[23] T. Floquet, C. Edwards, S.K. Spurgeon, On sliding mode observers for systems with unknown inputs, International Journal of Adaptive Control and Signal Processing 21 (8–9) (October–November 2007) 638–656.
[24] Z. Gao, S. Hu, F. Jiang, A novel motion control design approach based on active disturbance rejection, in: Proceedings of the 40th IEEE Conference on Decision and Control, Orlando, Florida, USA, December, 2001, pp. 4877–4882.
[25] D.W. Gu, F.W. Poon, A robust state observer scheme, IEEE Transactions on Automatic Control 46 (12) (Dic. 2001) 1958–1963.
[26] F. Gavilan, A. Acosta, R. Vazquez, Output-feedback control of the longitudinal flight dynamics using adaptative backstepping, in: Conference on Decision and Control and European Control Conference (CDC-ECC), Orlando, FL, USA, 12–15 Dec., 2011.
[27] M.S. Grewal, A.P. Andrews, Kalman Filtering: Theory and Practice Using MATLAB, 3rd edition, Wiley, USA, 2008.
[28] M. Garcia, T. Muskardin, Analysis and development of a reliable fixed wing UAV control system for mission profiles with restricted GPS availability, in: International Conference on Unmanned Aircraft Systems (ICUAS), Atlanta, GA, USA, May 28–31, 2013.
[29] M. Green, D.J.N. Limebeer, Linear Robust Control, Dover Publications, USA, 2012.
[30] F.J.J. Hermans, M.B. Zarrop, Sliding-mode observers for robust sensor monitoring, in: Proceedings of 13th IFAC World Congress, San Francisco, USA, 1996, pp. 211–216.
[31] J. Han, A class of extended state observers for uncertain systems, Control and Decision 10 (1) (1995) 85–88 (in Chinese).
[32] Y. Hou, Z. Gao, F. Jiang, B.T. Boulter, Active disturbance rejection control for web tension regulation, in: Proceedings of the 40th IEEE Conference on Decision and Control, Orlando, Florida, USA, December, 2001, pp. 4974–4979.
[33] T. Hu, Z. Lin, Control Systems with Actuator Saturation: Analysis and Design, Birkhaüser, USA, 2001.
[34] H. Wu, Adaptive stabilizing state feedback controllers of uncertain dynamical systems with multiple time delays, IEEE Transactions on Automatic Control 45 (9) (Sep. 2000) 1697–1701.
[35] H. Wu, Decentralized adaptive robust control for a class of large scale systems including delayed state perturbations in the interconnections, IEEE Transactions on Automatic Control 47 (10) (Oct. 2002) 1745–1751.
[36] H. Wu, Adaptive robust state observers for a class of uncertain nonlinear dynamical systems with delayed state perturbations, IEEE Transactions on Automatic Control 54 (6) (Jun. 2009) 1407–1412.
[37] J. Han, A class of extended state observers for uncertain systems, Control and Decision 10 (1) (1995) 85–88.
[38] H. Wassim, V. Chellaboina, Nonlinear Dynamical Systems and Control, Princeton University Press, Princeton, New Jersey, 2008.
[39] K. Jong-Hyuk, W. Stuart, S. Salah, Real-time navigation, guidance, and control of a UAV using low-cost sensors, in: 5th International Conference of Field and Service Robotics, Port Douglas, Australia, 29th–31st July, 2005.
[40] E.N. Johnson, S.K. Kannan, Nested saturation with guaranteed real poles, in: Proceedings of the American Control Conference, Denver, Colorado, USA, June 4–6, 2003.

[41] H.K. Khalil, Nonlinear Systems, third edition, Prentice Hall, USA, 1995.
[42] R. Kelly, V. Santibañez, Control de movimiento robots manipuladores, Prentice Hall, México, 2003.
[43] J. Kellery, D. Thakury, V. Dobrokhodov, K. Jones, Coordinated commencement of pre-planned routes for fixed-wing UAS starting from arbitrary locations a near real-time solution, in: International Conference on Unmanned Aircraft Systems (ICUAS), Orlando, FL, USA, May 27–30, 2014.
[44] M. Krstic, I. Kanellakopoulos, P. Kokotovic, Nonlinear and Adaptive Control Design, Wiley and Sons, USA, 1995.
[45] D. Luenberger, Observers for multivariable systems, IEEE Transactions on Automatic Control 11 (1966) 190–197.
[46] A. Levant, Robust exact differentiation via sliding mode technique, Automatica 34 (3) (1998) 379–384.
[47] A. Levant, Construction principles of 2-sliding mode design, Automatica 43 (4) (2007) 576–586.
[48] A. Levant, Higher-order sliding modes, differentiation and output-feedback control, International Journal of Control 76 (9–10) (2003) 924–941.
[49] J. Leyva, A.E. Pearson, An asymptotic modal observer for linear autonomous time lag systems, IEEE Transactions on Automatic Control 40 (7) (Jul. 1995) 1291–1294.
[50] D. Mclean, Automatic Flight Control Systems, Prentice Hall International, UK, 1990.
[51] D.R. Nelson, D.B. Barber, T.W. Mclain, R.W. Beard, Vector field path following for miniature air vehicles, IEEE Transactions on Robotics 23 (3) (2007) 519–529.
[52] K. Ogata, Modern Control Engineering, fifth edition, Prentice Hall, New Jersey, 2009.
[53] R. Sreedhar, B. Fernandez, G.Y. Masada, Robust fault detection in nonlinear systems using sliding-mode observers, in: Proceedings of IEEE Conference on Control Applications, Vancouver, BC, September, 1992, pp. 13–16.
[54] H. Sura-Ramírez, S.K. Spurgeon, A.S.I. Zinober, Sliding mode observer-controller design, in: Workshop on Robust Control via Variable Structure and Lyapunov Techniques (VSLT94), Benevento, Italy, September 7–9, 1994.
[55] A. Robertsson, R. Johansson, Observer backstepping and control design linear systems, in: Proceedings of the 37th IEEE Conference on Decision and Control, Tampa, Florida, USA, December, 1998, pp. 4592–4593.
[56] B.L. Stevens, F.L. Lewis, Aircraft Control and Simulation, John Wiley and Sons, USA, 1995.
[57] S. Sundaram, C.N. Hadjicostis, Delayed observers for linear systems with unknown inputs, IEEE Transactions on Automatic Control 52 (2) (Feb. 2007) 334–339.
[58] A. Seuret, T. Floquet, J.P. Richard, S.K. Spurgeon, A sliding mode observer for linear systems with unknown time varying delay, in: Proc. 2007 American Control Conf., New York, Jul., 2007, pp. 4558–4563.
[59] J.J. Slotine, J.K. Hendick, E.A. Misawa, On sliding observers for nonlinear systems, Journal of Dynamic Systems, Measurement, and Control 109 (1987) 245–252.
[60] Y. Shtessel, C. Edwards, L. Fridman, A. Levant, Sliding Mode Control and Observation, Birkhauser, London, UK, 2010.
[61] R. Trabelsi, A. Khedher, M. Mimouni, M. M'Sahli, An adaptive backstepping observer for on-line resistance application, International Journal of Sciences and Techniques of Automatic Control and Computer Engineering 4 (1) (July 2010) 1246–1267.
[62] K. Turkoglu, E.M. Jafarov, Augmented optimal LQR control system design for the longitudinal flight dynamics of an UAV: inner and outer loop concepts, in: Proceedings of the 9th WSEAS International Conference on Automatic Control, Modeling & Simulation, Wisconsin, USA, 27–29 May, 2007.

[63] A.R. Teel, Global stabilization and restricted tracking for multiple integrators with bounded controls, Systems & Control Letters 18 (3) (1992) 165–171.

[64] V.I. Utkin, Sliding-Modes in Control Optimization, Springer-Verlag, Berlin, Heidelberg, 1992.

[65] V. Grantham, Modern Control Systems Analysis and Design, John Wiley and Sons, London, 1993.

[66] H. Wu, Adaptive stabilizing state feedback controllers of uncertain dynamical systems with multiple time delays, IEEE Transactions on Automatic Control 45 (9) (Sep. 2000) 1697–1701.

[67] H. Wu, Decentralized adaptive robust control for a class of large scale systems including delayed state perturbations in the interconnections, IEEE Transactions on Automatic Control 47 (10) (Oct. 2002) 1745–1751.

[68] H. Wu, Adaptive robust state observers for a class of uncertain nonlinear dynamical systems with delayed state perturbations, IEEE Transactions on Automatic Control 54 (6) (Jun. 2009) 1407–1412.

[69] W. Wang, Z. Gao, A comparison study of advanced state observer design techniques, in: American Control Conference, vol. 6, Denver, Colorado, USA, December, 2003, pp. 4590–4595.

[70] B.L. Walcoll, S.H. Zack, State observation of nonlinear uncertain dynamical systems, IEEE Transactions on Automatic Control 32 (2) (Feb 1987) 166–170.

[71] Y. Li, C. Chen, W. Chen, Research on longitudinal control algorithm for flying UAV based on LQR technology, International Journal on Smart Sensing and Intelligent Systems 6 (5) (2013) 2155–2181.

Index

FSC
www.fsc.org
MIX
Papier | Fördert
gute Waldnutzung
FSC® C083411

Kolibri 360 GmbH
Ferdinand-Jühlke-Straße 7
99095 Erfurt, Deutschland
produktsicherheit@kolibri360.de